Electrostatics - Fundamentals and Modern Applications

Edited by Dengming Xiao and Su Zhao

Published in London, United Kingdom

Electrostatics - Fundamentals and Modern Applications
http://dx.doi.org/10.5772/intechopen.111059
Edited by Dengming Xiao and Su Zhao

Contributors
Abdallah Alkhaiyat, Dandan Qu, Fangming Ruan, Galina Mratskova, Kun Zhang, Manisha Muduli, Michal Sahul, Mohamed Badran, Peter Janiga, Qingrong Peng, Sri Harsha Kodati, Toshikazu Shinba, Xing Fan, Yang Meng, Yan Lyu, Ze Chen, Zhehao Pei

First published in London, United Kingdom, 2025 by IntechOpen
IntechOpen is the global imprint of INTECHOPEN LIMITED, registered in England and Wales, registration number: 11086078, 167-169 Great Portland Street, London, W1W 5PF, United Kingdom

For EU product safety concerns: IN TECH d.o.o., Prolaz Marije Krucifikse Kozulić 3, 51000 Rijeka, Croatia, info@intechopen.com or visit our website at intechopen.com.

British Library Cataloguing-in-Publication Data
A catalogue record for this book is available from the British Library

Electrostatics - Fundamentals and Modern Applications
Edited by Dengming Xiao and Su Zhao
p. cm.
Print ISBN 978-1-83769-476-1
Online ISBN 978-1-83769-475-4
eBook (PDF) ISBN 978-1-83769-477-8

If disposing of this product, please recycle the paper responsibly.

Meet the editors

Dengming Xiao was born in Guangxi Province, China, in 1953. He received his Ph.D. degree from the School of Electrical Engineering, Xi'an Jiaotong University, China, in 1994. Currently, he is a Professor at the Department of Electrical Engineering at Shanghai Jiao Tong University, Shanghai, China. He has worked on gas insulation systems, electrical equipment online monitoring and fault diagnosis for more than 30 years.

Su Zhao is a research assistant in the Department of Electrical Engineering at Shanghai Jiao Tong University, China. He received his BSc degree from Xi'an Jiaotong University in 2014 and his Ph.D. from Shanghai Jiao Tong University in 2019. He was a postdoctoral researcher at the University of Liverpool, UK, from 2019 to 2021. His primary research interests include gas discharge and gas insulation.

Contents

Preface

Electrostatics, a cornerstone of classical physics, has long fascinated scientists and engineers with its elegant principles and profound implications. From the early experiments of Coulomb to the sophisticated applications in modern technology, the study of electric charges at rest has continually evolved, revealing new insights and enabling groundbreaking innovations. This book, *Electrostatics – Fundamentals and Modern Applications*, seeks to bridge the gap between the foundational theories of electrostatics and their cutting-edge applications in today's world.

As an editor and a professor deeply immersed in the field, I have witnessed the transformative impact of electrostatics across diverse disciplines. This book is designed to serve as a comprehensive resource for students, researchers, and professionals who seek to deepen their understanding of electrostatics and explore its relevance in contemporary science and engineering. The chapters are carefully curated to provide a balanced blend of theoretical rigor and practical relevance, ensuring that readers gain both a solid conceptual foundation and an appreciation for the real-world applications of electrostatics.

This project would not have been possible without the contributions of a talented group of authors, each of whom brings a unique perspective and expertise to the subject. I am deeply grateful for their dedication and intellectual rigor, which have been instrumental in bringing this book to fruition. I would also like to extend my gratitude to IntechOpen for their support and commitment to advancing scientific knowledge through open-access publishing. Their vision aligns perfectly with the goal of this book: to make the fascinating world of electrostatics accessible to a global audience.

Finally, I hope this book will inspire readers to appreciate the beauty of electrostatics and its profound impact on our understanding of the physical world. Whether you are a student embarking on your first course in electromagnetism, a researcher exploring new frontiers, or a professional seeking to apply electrostatic principles in your work, I trust that this book will serve as a valuable guide and a source of inspiration.

Welcome to the world of electrostatics—a world where fundamental principles meet modern innovation and where the mysteries of electric charges continue to spark curiosity and discovery.

Dengming Xiao
Department of Electrical Engineering,
Shanghai Jiao Tong University,
Shanghai, China

Su Zhao
Shanghai Jiao Tong University,
Shanghai, China

Chapter 1

Application of the Low Frequency and Low-Intensity Electrostatic Field in Rehabilitation Programs for Patients with Musculoskeletal Disorders

Galina Mratskova

Abstract

A low-frequency and low-intensity pulsed electrostatic field is applied as a treatment method in Rehabilitation Medicine. The technique is known as Deep Oscillation® therapy. At present, it has been established that the therapeutic effects of the electrostatic field on the tissues depend on the field frequency (5–250 Hz). The advantage of the method is the operation with a low amperage of 5–7 μA and a bipolar form of the wave. Deep Oscillation® was originally created to improve tissue lymphatic drainage. Subsequently, it was found that it reduces acute and chronic pain, improves blood and lymph circulation in the region of application, and reduces edema and fibrosis. In addition to this, tissue elasticity, joint mobility, and the condition of muscle and nerve tissue are improved. The aim of this research project is to investigate the therapeutic impact of a low-frequency and low-intensity pulsed electrostatic field in patients with symptomatic knee osteoarthritis.

Keywords: low-frequency and low-intensity pulsed electrostatic field, Deep Oscillation® therapy, symptomatic knee osteoarthritis, rehabilitation, pain, edema, range of motion, functional activity

1. Introduction

The therapeutic impact of the Deep Oscillation® modality has been applied in Physical and Rehabilitation Medicine recently. This method uses Low-frequency and Low-intensity electrostatic field, which causes a gently acting vibration (oscillation) deep in the tissues. This electrostatic phenomenon is explained through the Johnson-Rahbeck effect. In this phenomenon, there is a dependence between the applied voltage and the force of attraction and relaxation between the contact materials (metal surface and semiconductor material) when an electric potential is applied at the boundary between them [1]. For the needs of physical therapy, an electrostatic field is created that impacts the tissues (skin, muscles, blood and lymphatic vessels, nervous structures, and connective tissue) that are located below the source creating oscillation [2–4].

The first medical device using a Low-frequency and Low-intensity electrostatic field was developed by H. Seidel и W. Waldner at the end of the twentieth century [4]. Deep Oscillation® therapy is a patent-enhanced practical therapy method that works on the basis of the pulsed electrostatic field of low intensity (U = 100–400 V; I = 150 μA) [2] and low frequency (5–250 Hz). As a result of its application on the surface of the skin in the depth of the application area, a vibration (oscillation) is formed in the underlying biological tissues [2, 3, 5–7].

1.1 Deep Oscillation® therapy - Characteristic of low frequency and low intensity electrostatic field treatment

The following therapeutic effects have been reported from clinical studies conducted up to date:

- Reduction of pain (acute or chronic) in various clinically manifested diseases [2, 5, 6, 8–17];
- Anti-inflammatory effect [18, 19];
- Affects lymphedema with primary or secondary genesis [2, 5, 20–23];
- Reduces and modulates the process of tissue fibrosis [24, 25];
- Leads to reduction of increased muscle tone and relaxation, mobilizes fascial tissue, and increases reduced range of motion in joints [2, 8, 9, 26–28];
- Improves and accelerates the tissue recovery process in wound defects [12, 29];
- It improves the process of recovery after a training cycle, decongestion, and leads to improvement of the function of the respiratory system in obstructive lung diseases [12, 30–32].

The therapeutic effect of Deep Oscillation® is based on the pulsating electrostatic field of very low frequency and low intensity, an electromagnetic field in which no thermal effect is observed [7, 10, 33]. The mechanism of physiological action is carried out in the formation of tissue vibration to a depth of 8 cm (skin, subcutaneous tissue, connective tissue, neuromuscular apparatus, blood, and lymph vessels), which are located in the area under the applicator during its movement in the sagittal plane [7].

The frequency of the electrostatic field is low and varies from 5 to 250 Hz. The established physiological effects are directly dependent and determined by the applied frequency. At a frequency in the range of 80–250 Hz, an analgesic, spasm-reducing, and anti-edematous effect is observed; the frequency in the range between 25 and 80 Hz leads to the improvement of the metabolism and recovery processes in the tissues, the venous and lymphatic circulation is optimized; frequency in the range of 5–25 Hz can improve local hemodynamics, the functional status of muscle tissue, and tissue trophicity. For the needs of practice, a small current amperage (5–7 μA) is used. The pulses have a biphasic form, and the usual electrotherapy effects of electrolysis are not observed. The electrostatic field is discharged during each pause, which protects the patient's body from the accumulation of static electricity [7, 10].

There are two main methods of practical application of Deep Oscillation® therapy. One method uses a handheld applicator (oscillating head) with a different size in diameter depending on the area of the field being treated. In the other application method, the therapist's hands are connected to the oscillation device and simultaneously placed in vinyl gloves. The therapist performs the basic massage techniques, which allows to combine the effect of the electrostatic field with massage. In this second method, the electrostatic field affects both the therapist and the patient at the same time, which implies that the rules of contraindications must be strictly observed for both at the same time. The therapeutic course includes between 5 and 15 procedures, which can be carried out every day or every other day. Deep Oscillation® can be conducted in combination with therapeutic exercises, cryo or heat therapy, as well as other physical modalities (electrotherapy). To achieve a better effect, it is recommended to sequentially apply combinations of different modalities with different durations depending on the phase of the disease in one procedure.

Deep Oscillation® therapy can be a suitable therapeutic alternative for various diseases including:

- Traumatic, degenerative, and inflammatory diseases of the musculoskeletal system [10, 34–39];

- In the recovery period: after bone fracture, after surgical interventions on different occasions, and after osteotomy, metal osteosynthesis, joint replacement, to prevent the appearance of decubitus wounds and the formation of rough cicatricial tissue [15, 19, 40–42];

- In some diseases of the nervous system of peripheral origin (dorsopathies, etc.) [33];

- After pathological damage of central nervous origin (damage to the brain and spinal cord as a result of vascular or post-traumatic damage to the brain and spinal cord, Parkinson's disease, multiple sclerosis, and migraine [7, 43–45];

- In cardiovascular diseases (hypertension, chronic diseases occurring with insufficiency of the venous and lymphatic systems, lymphedema, etc.) [2, 7, 21];

- In pathology of the respiratory system, which includes inflammatory diseases (chronic bronchitis, pneumonia), bronchial asthma, COPD, and cystic fibrosis [32];

- In some diseases of the digestive tract (chronic gastritis or gastro duodenitis), diseases with disturbed motility of the bile ducts; in inflammatory diseases of the upper respiratory tract–chronic sinusitis, laryngitis [7];

- After extracorporeal lithotripsy in renal calculus, chronic nonbacterial prostatitis [7];

- For the treatment of wounds of various origins and burns [7, 29].

Deep Oscillation® therapy provides a number of advantages, which makes it a preferred physiotherapeutic method for acute and chronic diseases, both of the musculoskeletal system that cause pain, and for diseases of the central and peripheral nervous system, after acute, chronic, and sports injuries, for the prevention of thrombotic phenomena. The inclusion of the intervention with Deep Oscillation® shortens the time of hospital treatment, reduces the period required for recovery, improves lymphatic and venous circulation, and has a lymphatic drainage effect, which is a consequence of the gentle impact on the target tissues. The therapy leads to relaxation of the muscles that have increased tone. Treatment with the Deep Oscillation® device is pleasant for the patient and comfortable and easy for therapists to use. Despite the many benefits and few side effects, with Deep Oscillation® therapy, it is necessary to monitor for manifestations of a sense of sharp pain in the area of application, increased sensitivity during urination, symptoms of fatigue and/or exhaustion, increased local temperature, and signs of hypotension [10].

1.2 Characteristics of knee osteoarthritis

Osteoarthritis (OA) is a disease of the musculoskeletal system with a widespread occurrence worldwide. It can affect any of the upper/lower limb and axial skeleton joints, but degenerative changes are most common in the weight-bearing joints of the lower limbs with the knee joint (KJ) being most affected [46]. The disease is linked to a negative impact and increased economic burden, both on the part of patients suffering from OA and on the part of society, due to increased treatment costs and reduced work capacity [47]. Knee OA has the highest proportion (over 80%) in the distribution of disease burden. According to US research data, approximately 19% of the population over 45 years of age exhibit clinical symptoms of OA [48–51].

Osteoarthritis is typically more prevalent after the sixth decade, in individuals 65 years of age and older [50, 52, 53]. The disease impairs the joint structurally, including destruction and reduction of cartilage tissue, which is demonstrated by limitations in functional activity [54]. Routine radiological images show a narrowed interarticular gap and subchondral bone changes, with gradual appearance of osteophytes. In practice, however, joint pain sensations are often poorly correlated with the changes seen in joint structure [55]. Knee pain is not reported by a significant proportion of patients whose imaging studies demonstrate the presence of OA [56, 57].

Symptomatic OA is thought to be preceded by a prolonged preclinical phase characterized by the appearance and development of structural changes in the joint without the presence of clinical symptoms and subjective complaints. Not only changes in the subchondral bone and articular cartilage but also changes in the muscles around the joint are essential. Muscle weakness is one of the earliest and most common symptoms in patients with knee osteoarthritis and is considered a better predictor of joint narrowing and pain [55, 58]. Decreased functional activity and disability have a significant impact on the psychosocial status of both patients and their families [59, 60]. In recent years, the severity of musculoskeletal diseases has increased Wallace et al. [51] believe that the prevalence of knee OA is 2.1 times higher in postindustrial society than in the early industrialization era [51]. There is a global trend of increasing life expectancy, which suggests that the need for health care for these patients will increase. Until recently, population aging and high body mass index were considered to be major factors in the prevalence of OA [51], but the role of other factors is likely to need to be assessed.

According to some investigators, symptomatic osteoarthritis of the knee with clinical symptoms is present in only about 15% of those affected with radiological evidence of OA [53, 61]. This is probably due to the slow evolution of the disease, which is preceded by a prolonged asymptomatic phase during which, in the absence of clinical symptoms and history complaints, degenerative changes slowly progress. The changes affect not only the subchondral part of the bone and the cartilage plate but also the joint motor muscles. One of the first signs commonly seen in knee OA is weakness of the surrounding joint musculature, and it is even considered to be a better prognostic sign than a decrease in joint space dimensions and pain sensation [55, 58]. The pathological changes occurring in the OA joint lead to increased psycho-emotional strain in patients and their families due to decreased functional activity and associated negative effects [59, 60]. In relation to the increase in degenerative musculoskeletal diseases and their burden in modern society, Wallace et al. [51] assumed that the prevalence of knee osteoarthritis is twice as high today compared to the early industrialization period [51]. According to objective data, life expectancy is increasing in countries around the world, which in turn will lead to increased demand for healthcare services. On the other hand, a reassessment of the major risk factors and their role in the prevalence of OA will probably have to be made, although today, the increase in the elderly population and the high body mass index are considered to be the main ones [51].

Recurrent inflammatory responses derived from excessive mechanical loading of the joints [62, 63], abnormal weight gain, gender, hormone levels, age, and possible past trauma have a direct impact on the onset and development of OA and are considered to be factors of high significance. Concurrent with these, gradual changes in the surrounding musculature, including that of the knee, are also likely to be important in the onset and progression of the disease. Decreases in muscle strength is likely to be a uniting factor in the interaction between significant risk factors [55, 64, 65]. Gradually, chronicifying pain and progressive functional impairment lead to an increased need for medical care and rehabilitation, which in turn increases healthcare costs. Therefore, it is important that the patients form an active and responsible attitude regarding their own health in accordance with the current recommendations for the treatment of musculoskeletal degenerative diseases. Self-monitoring, personal healthcare, and active collaboration with health professionals are important elements of the OA self-management process. Good awareness of the positive effects of lifestyle changes, physical activity, and possible therapeutic approaches in the treatment of OA are also likely to improve health outcomes [66]. Despite the availability of various treatment programs for OA, there is an ongoing search for resources to develop long-term programs that will be effective in the treatment and rehabilitation of musculoskeletal disorders directly related to the joint degeneration process [67, 68].

2. Therapeutic use of Deep Oscillation® therapy in the rehabilitation program of patients with knee osteoarthritis

The aim of the research project is to investigate the therapeutic effects of the low-frequency and low-intensity pulsed electrostatic field application, using the Deep Oscillation®-therapy in patients with knee osteoarthritis and to compare the therapeutic effects of the application of Deep Oscillation® and placebo-Deep Oscillation® and kinesitherapy.

2.1 Materials and methods

The study included 90 patients with knee OA, with symptomatic osteoarthritis of the knee joint in a subacute and chronic phase without clinical evidence stage of active inflammation, randomized into two groups: Therapeutic (n = 57) and Control group (n = 33).

Treatment is carried out in accordance with the Helsinki Declaration (1964), after informed consent. The rehabilitation program included a ten-day therapeutic course: Deep Oscillation® and placebo-DO and Kinesitherapy (KT) complex.

Patients were assessed at four time points before, after treatment, after 1 month, and 3 months after therapy, using a visual analog scale (VAS) to measure pain at rest, at move, climbing and descending stairs, Manual Muscle Test, Measurement of the knee joint circumference, Test Range of Motion and WOMAC Osteoarthritis Index (Western Ontario and McMaster University Osteoarthritis Index).

2.2 Criteria for inclusion and exclusion

2.2.1 Criteria for inclusion

The research included eligible patients who satisfied the following criteria: age more than 38 years; X-ray data for knee OA of Kellgren—Laurence grades II and III; pain in knee joint in most days of recent months; crepitations during active movement of the knee joint; morning stiffness lasting 30 minutes or less.

2.2.2 Criteria for exclusion

Age less than 38 years; X-ray data for knee OA of Kellgren—Laurence grade I and IV; data on OA accompanied by a history of acute knee injury, reflected pain from the lumbar spine or hip joint; active synovitis; treatment with nonsteroidal anti-inflammatory drugs NSAIDs and / or analgesics, intraarticular manipulations with hyaluronic acid, or corticosteroids up to 6 months after application; contraindications to electrotherapy, including pacemaker.

2.3 Characteristics of the therapeutic course by groups

The therapy is administered in 10 sessions. All Therapeutic group patients underwent a therapeutic course with low-frequency and low-intensity electrostatic field and complex of therapeutic exercises and Placebo-Deep Oscillation and complex of therapeutic exercises for Control group. All patients included in the study gave informed consent to participate in the treatment course and could give up at any time from the treatment course.

2.3.1 Therapeutic group (TG)

In the Therapeutic group (TG), the low frequency, and low-intensity electrostatic field was applied by the Deep Oscillation® method using a 9.5 cm hand applicator (Deep Oscillation Evident® CLINICS, Physiomed Elektromedizin AG). The procedure involves treatment of KJ, surrounding tissues, and the m. Quadriceps femoris area. Therapeutic Modality: Variable Frequency 100–144 Hz 5 min. Permanent frequency 85 Hz 5 min.

Variable frequency 14–20 Hz 4 min. Modulation mode 1:1 (light vibrations). The complex of therapeutic exercises was carried out immediately after the electrical procedure.

2.3.2 Control group (CG)

Control group (CG) Patients underwent a ten-day placebo course using the Deep Oscillation® method. The application methodology was similar to the Therapeutic group, with minimal intensity (device turned on, contact indicator activated) but no deep tissue vibration in the treated area. The placebo procedure is followed by the therapeutic exercise complex.

The therapeutic exercise program included aerobic exercise, analytical exercise for femur muscles, focusing on Vastus medialis et lateralis m Quadriceps femoris, relaxation techniques for shortened muscles, resistance exercises, and exercises to increase the range of motion in the knee joint.

The possible therapeutic goals of the rehabilitation of patients with symptomatic gonarthritis with low-frequency and low-intensity electrostatic field and complex therapeutic exercises were reduction of pain and stiffness, reduction of edema of the around joint tissues and muscular weakness, increasing range of motion in affected joint and locomotive function, and increase of the daily functional activity.

2.4 Tools for evaluating the studied indicators

2.4.1 Functional methods for kinesiological assessment of the knee joint

2.4.1.1 Manual muscle testing (MMT)

Manual Muscle Testing (MMT) is a method used to determine the degree of muscle weakness. Skeletal muscles can have weakness as a result of disease, trauma, or reduced or lack of movement. Testing on the MMT scale provides information that allows the creation of an exact program of therapeutic exercises or other physical methods for disorders in the musculoskeletal system. The scale enables muscle weakness to be assessed in six grades (0–5 grade). 5, 4, and 3 according to MMT are accepted as functional grades. There are cases where these grades do not accurately reflect muscle weakness. In these circumstances, a„ + "or„ – "sign is added to the corresponding test item, corresponding to 5 to 10% strength. 3 (+) corresponds to a movement that can be performed and repeated several times against gravity or once against light resistance; 3 (−) corresponds to movement that can overcome gravity but has an incomplete range of 50–90% of the range of normal movement; score 2 (+) – corresponds to initiated antigravity movement (50% or less of normal range of motion); A score of 2 (−) receives movement performed with gravity eliminated and at incomplete range [11].

In order to objectively present the results of the performed MMT testing, when the signs "+" or "-" are used to the corresponding degree, numerical equivalents for determining muscle weakness are used, which are presented in **Table 1**.

MMT	2	2+	3-	3	3+	4-	4	4+	5-	5
	2	2.25	2.75	3	3.25	3.75	4	4.25	4.75	5

Table 1.
Degrees of muscle weakness according to the Manual muscle test (MMT).

2.4.1.2 Range of motion

In this kinesiological measurement, joint plane range of motion (SFTR) is determined using a goniometer. It is one of the main parameters characterizing motor function. For functional assessment of the knee joint, the measurement is in the sagittal plane, in which the range of extension and flexion in the tested joint is considered: S – 0°-0°-130°.

2.4.1.3 Measurement of circumferences

Centimetry is a kinesiological method for measuring anthropometric indicators. With a plastic centimeter, the condition of the musculature is determined and the presence of pathological conditions, such as muscle hypotrophy or atrophy, joint circumference—in different conditions, norm or pathology (edema) [11].

2.4.2 Pain and functional activity self-report questionnaires for knee joint assessment

2.4.2.1 Visual analogue scale (VAS)

In practice, this scale is routinely used to assess the degree of pain in different age groups and in patients with different diseases. VAS is a one-dimensional scale, representing a straight line from 0 to 10 cm (0 mm to 100 mm). At the beginning, a baseline value of 0, which corresponds to a no-pain condition, is noted. At the end, the maximum value corresponding to the worst possible pain is noted. Between these two values, there are intermediate values that can describe the nuances in subjective pain (2—corresponds to mild pain, 4- indicates moderate pain, 6- corresponds to severe pain, 8- is indicative of very severe pain). It is necessary for patients to independently determine the level of their pain by marking on the line how strong a feeling of pain they have.

Knee joint pain was assessed using a Visual Analogue Scale (VAS), assessing pain at rest, walking, descending, and climbing stairs.

2.4.2.2 Western Ontario and McMaster universities osteoarthritis index (WOMAC index)

A self-administered WOMAC Osteoarthritis questionnaire was used to assess subjective complaints and degree of functional impairment. It was developed for adult patients with damage to the knee and hip joints. Osteoarthritis Index LK 3.1 modification was used in the study.

The WOMAC Index is a disease-specific for knee and hip OA, a self-administered questionnaire used for subjective self-assessment of health and physical function.

It consists of 24 elements, divided into three subscales: Pain (5 questions), Stiffness (2 questions), and Physical function (17 questions).

The severity of symptoms is assessed on the verbal Likert scale in 5 points: 0—None, 1—Mild, 2—Moderate, 3—Severe, 4—Extreme.

Higher scores on the WOMAC Index indicate worse pain, stiffness, and functional limitations. The results are generated by summing the coded responses. Psychometric studies show high validity and reliability of the WOMAC questionnaire.

2.5 Evaluation of the obtained results

Results were tracked at the beginning, immediately after therapy, on the 1st month after rehabilitation and after the 3rd month of the end of the treatment course by: Visual Analogue Scale (VAS) pain test at rest, at move, climbing and descending stairs, Manual Muscle Test, Measurement of the knee joint circumference, Test Range of Motion, and WOMAC Osteoarthritis Index (Western Ontario and McMaster University Osteoarthritis Index), developed for adult patients with osteoarthritis. Modification Osteoarthritis Index LK 3.1 has been used. Statistical analysis: Quantitative data is represented as Mean and Standard Deviation (SD) if the distribution is either normal or Median (Me) and Range if different than normal. The normality was tested by the Shapiro-Wilk test. Comparisons of quantitative data before and after therapy were performed by the One sample Student t-test or the Wilcoxon Signed Ranks test. The category data was presented in percentages and analyzed by Fisher's Exact Test. The measured WOMAC Index in patients is presented as Mean ± SD (Range) standard deviation (range). The difference between a pair of groups was assessed using the nonparametric Mann-Whitney U test. Changes in the various WOMAC scales over the four time periods

Indicators	**Therapeutic group (n = 57)**	**Control group (n = 33)**
Age (Mean ± (SD)Range)	65.6 ± 11.3 (42–85)	67.6 ± 9.5 (44–87)
Gender identity (Mean ± (SD)Range)	42 women 65.9 ± 11.7(42–85)	22 women 65.9 ± 11.7(42–85)
	15 men 64.9 ± 10.4(45–81)	15 men 65.9 ± 8.4(48–76)
Duration of the disease (Me(Range)) years	10(1–15) years	7(2–15)
Duration of the current exacerbation period (Me(Range)) weeks	6.37 ± 1.74 (4–12)	6.27 ± 1.88 (3–10)
X-ray degree of Kellgren-Lawrence scale:		
II degree n (%)	30 (52.6)	20 (60.6%)
III degree n (%)	27 (47.4)	13 (39.4)
Body weight in kg.	78.0 (50–155)	82.0 (60.0–91.0)
BMI kg/cm^2	29.1 ± 5.4 (20.0–58.3)	28.6 ± 2.9 (23.3–35.1)
Reason for visiting a doctor (the number (%) is greater than the total (100%) for the group because some patients donated more than one answer)		
Pain	57 (100.0)	33 (100.0)
Difficulties while walking	47 (82.5)	28 (84.8)
Stiffness	22 (38.6)	11 (33.3)
Limited daily activity	16 (28.1)	10 (30.3)

Table 2.
Characteristics of the patients in therapeutic and control group.

were analyzed using the Friedman test. The doubles comparison in the presence of interdependence across the periods was made with a Dunn *post hoc* test. Data were analyzed with SPSS for Windows v.24. Significance level $p < 0.05$ was used to determine the statistical differences.

2.6 Characteristics of the patients in therapeutic and control group

90 patients with clinical symptoms, diagnosed with II and III stage of gonarthritis according to Kellgren-Lawrence, aged between 42 and 85 years, were included in the study. Patients were randomized into two groups: Therapeutic group ($n = 57$) and Control group ($n = 33$). Characteristics of the patients in Therapeutic and Control groups are shown in **Table 2**.

3. Therapeutic results of Deep Oscillation therapy in the rehabilitation program of patients with knee osteoarthritis

3.1 Comparative analysis of data in MMT m. Quadriceps femoris by groups

After the therapeutic course in the Therapeutic group, there was a statistical difference in MMT values between the four follow-up periods ($p < 0.001$). Muscle weakness decreased significantly immediately after completion of Rehabilitation, 1st month ($p < 0.001$), and 3rd month ($p < 0.001$) compared to the beginning of therapy. The treatment effect was maintained between months 1 and 3 ($p = 0.822$) (**Table 3**).

In the Control group, there was a decrease in muscle weakness compared to baseline at two time points (after completion of the treatment course ($p < 0.001$) and 1 month of therapy ($p < 0.001$). At month 3 of treatment, no statistical difference was found ($p = 0.192$) and MMT values returned to pre-therapy values. The effect of therapy persisted up to 1 month, compared to after treatment ($p = 1.000$). At month 3, there was an increase in muscle weakness compared to posttreatment as well as compared to month 1 ($p < 0.001$) (**Table 3**).

3.2 Comparative analysis for measured knee joint flexion range by group

The mean value of flexion range (Mean ± SD (Me (range)) before the start of the rehabilitation course in the Therapeutic group was 106.4 ± 8.3(110° (90.0–120.0)° and for the control group was 108.6 ± 8.4 (110°(90.0–125.0)°. For Therapeutic group, there was statistically significant difference in flexion range before therapy versus after therapy ($p < 0.001$), at 1st month ($p < 0.001$), and at 3rd month ($p < 0.001$). There was no statistical difference between flexion measured after the 1st and 3rd

Periods	Therapeutic group	Control group
Before therapy	3.00 (2.25–4.0)	3.00 (2.72–4.0)
After therapy	3.75 (3.00–4.75)	3.75 (3.00–4.75)
After 1st month	4.00 (3.25–5.00)	4.00 (2.75–5.00)
After 3rd month	4.00 (3.25–5.00)	3.25 (3.00–4.25)

Table 3.
Dynamics in MMT testing—m. Quadriceps femoris (Me (range)) for the four time points by group.

months ($p = 1.000$). A prolonged therapeutic effect was achieved up to 3 months after rehabilitation (**Table 4**).

3.3 Comparative analysis of knee joint measurement of circumferences by groups

The measured knee joint circumferences in the groups were followed at the beginning of the therapeutic course, immediately after therapy at 1 and 3 months. Mean ± SD (Me (range)) Knee joint centimetry before the start of the rehabilitation course in the Therapeutic group was 40.9 ± 4.0 (40.0(32.0–52.0) cm, control group was 41.3 ± 2.3(42.0(36.0–44.5))cm (**Table 5**).

For the Therapeutic group ($n = 57$), there was a statistically significant difference in the measurement of circumferences values before therapy compared to after therapy ($p < 0.001$), at 1st ($p < 0.001$), and at 3rd months ($p < 0.001$). There was no statistical difference in measurement of circumferences values after therapy and at 1st month ($p = 1.000$), and after therapy and at 3rd month ($p = 1.000$) and between 1st month and 3rd month ($p = 0.162$). A sustained therapeutic effect was achieved up to 12 weeks after rehabilitation (**Table 5**).

For the Control group ($n = 33$), there was a statistical difference in measurement of circumferences at month 3 compared to baseline ($p = 0.022$), after therapy and at month 3 ($p = 0.003$), and between month 1 and month 3 ($p = 0.034$). There was no difference in measurement of circumferences before and after therapy ($p = 1.000$) and at 1st month ($p = 1.000$), and after therapy and month 1 ($p = 1.000$).

3.4 Comparative analysis of knee joint pain (VAS) by groups

Pain in the affected knee joint was assessed by VAS before therapy, after therapy at months 1 and 3 for each treatment group in terms of pain at rest, while walking on flat surfaces, descending, and ascending stairs for the four time points. No statistically significant differences regarding gender and age distribution between groups were found in the statistical analyses performed.

Periods	Therapeutic group	Control group
Before therapy	106.4 ± 8.3(90.0–120.0)	108.6 ± 8.4(90.0–125.0)
After therapy	116.8 ± 6.3(100.0–125.0)	117.4 ± 7.3(100.0–130.0)
After 1st month	119.7 ± 4.7(110.0–125.0)	117.4 ± 6.3(105.0–125.0)
After 3rd month	120.3 ± 3.9(110.0–125.0)	113.3 ± 5.6(100.0–120.0)

Table 4.
Dynamics in knee joint flexion range KJ by group at the four time points (Mean ± SD (Me (range)).

Periods	Therapeutic group	Control group
Before therapy	40.9 ± 4.0 (32.0–52.0)	41.3 ± 2.3 (36.0–44.5)
After therapy	39.3 ± 3.7 (30.5–50.0)	41.2 ± 2.2 (36.0–44.0)
After 1st month	39.1 ± 3.7 (30.0–49.0)	41.4 ± 2.2 (36.0–44.0)
After 3rd month	39.4 ± 3.7 (30.0–49.0)	41.7 ± 2.2 (36.0–45.0)

Table 5.
Knee joint measurement of circumferences (Mean ± SD (Me (range)) for the four time points by group.

3.4.1 Pain at rest

Prior to therapy, there was no statistically significant difference in baseline VAS (Mean ± SD) rest pain scores between the Therapeutic group 2.72 ± 0.90 and the Control group 2.52 ± 0.71.

For the Therapeutic group, there was a statistically significant reduction in pain at rest according to VAS at the end of therapy ($p < 0.001$), 1st month ($p < 0.001$) and 3rd month ($p < 0.001$) compared to baseline values. It was also observed retention of pain level at rest at 1st month ($p = 0.082$) and 3rd month ($p = 0.490$) compared to post-therapy was observed (**Table 6**).

In the Control group, after completing the course of treatment, VAS pain decreased at 1st and 3rd months compared to baseline. Pain at rest increased after the 1st month, with the effect persisting until the end of the 1st month ($p = 0.002$) and such increasing at 3rd month compared to the 1st month post treatment ($p < 0.001$). With a more long-lasting effect in terms of pain at rest, treatment in TG showed no significant difference in pain level after treatment at 1st and 3rd months, and between 1st and 3rd months; that is, the effect of treatment persisted even 12 weeks after therapy. In CG, a significant reduction in pain was achieved by the end of the first month with a subsequent increase in pain at month 3rd compared to month 1st, but the pain level was lower compared to baseline values.

3.4.2 Pain while walking (Mean ± SD)

Pain while walking (Mean ± SD) before therapy for the Therapeutic group was 4.19 ± 1.01 and for the Control group - 3.85 ± 0.76. There was no statistically significant difference in baseline values between the groups.

The results of the Deep Oscillation therapy conducted in TG showed sustained reduction in pain while walking by the end of month 3rd compared to baseline values ($p < 0.001$) and retention of therapeutic effect and pain level to values achieved post therapy. No statistical difference was observed between months 1st and 3rd ($p = 0.386$).

In CG, pain while walking decreased after the end of therapy. The effect persisted until the end of the first month; then, there was a statistical difference between the end of therapy and month 3rd, and between month 1st and 3rd and an increase in pain while walking after month 1st ($p < 0.001$) (**Table 7**).

Longer-lasting effect in terms of pain while walking was the therapy carried out in TG, in which no statistical difference was observed after the end of the treatment compared to months 1st and 3rd; a significant reduction in pain syndrome was observed at months 1 as well as 3. In CG, the effect persisted up to 1st month, after which there was a statistical difference between posttreatment pain and 3rd month, as

Periods	Therapeutic group	Control group
Before therapy	2.72 ± 0.90	2.52 ± 0.71
After therapy	0.67 ± 0.83	0.82 ± 0.81
After 1st month	0.12 ± 0.33	0.45 ± 0.62
After 3rd month	0.23 ± 0.50	1.76 ± 0.75

Table 6.
Pain dynamics at rest (Mean ± SD (Me(range))).

Periods	Therapeutic group	Control group
Before therapy	4.19 ± 1.01	3.85 ± 0.76
After therapy	2.09 ± 0.74	2.15 ± 0.57
After 1st month	1.21 ± 0.62	1.73 ± 0.72
After 3rd month	1.51 ± 0.69	3.06 ± 0.66

Table 7.
Pain at walking dynamics (Mean ± SD (Me (range))).

well as between 1st and 3rd months, and an increase in the degree of pain on walking at 3rd versus 1st month (**Table 7**).

3.4.3 Pain when descending stairs (Mean ± SD)

There was no statistically significant difference between Therapeutic group and Control group with respect to baseline values. For Therapeutic group, there was a statistically significant improvement in pain when descending stairs at all four time points ($p < 0.001$), at the end of therapy, at 1st and 3rd months, compared to baseline values ($p < 0.001$), at 1st month ($p < 0.001$) and 3rd months ($p < 0.001$). Retention of pain level when descending stairs at month 3rd ($p = 1.000$), versus after the end of therapy (**Table 8**).

For Control group, there was a reduction in pain when descending stairs over the posttreatment period, at 1st and 3rd months ($p < 0.001$); an increase in the degree of pain when descending stairs after 1st month, with the effect persisting until the end of 1st month; and an increase in pain at month 3 ($p = 0.016$), compared to post therapy, and at month 3 compared to month 1 ($p < 0.001$) (**Table 8**).

After analyzing the results for pain when descending stairs, it was found that the treatment conducted in the Therapeutic group had a more long-lasting effect, with no statistical difference observed after therapy compared to months 1 and 3, and between months 1 and 3; that is, the effect of the treatment persisted at month 3 after therapy.

3.4.4 Pain when ascending stairs

The pain score (Mean ± SD) on VAS before therapy for TG was 5.68 ± 1.07 and CG 5.36 ± 1.06. There was no statistically significant difference in baseline values of this parameter between groups. The treatment performed in TG was effective in terms of pain reduction ($p < 0.001$). No statistical difference was observed for the posttreatment period and month 3 ($p = 0.001$); the posttreatment effect was maintained at month 3, but an increase in pain was found between the end of months 1 and 3 ($p = 0.031$).

Periods	Therapeutic group	Control group
Before therapy	7.05 ± 1.04	6.76 ± 0.90
After therapy	3.89 ± 0.94	4.27 ± 0.84
After 1st month	3.26 ± 0.64	3.94 ± 0.83
After 3rd month	3.61 ± 0.62	5.24 ± 0.97

Table 8.
Dynamics of pain when descending stairs (Mean ± SD).

Periods	Therapeutic group	Control group
Before therapy	5.68 ± 1.07	5.36 ± 1.06
After therapy	2.96 ± 0.82	3.09 ± 0.88
After 1st month	2.40 ± 0.70	2.73 ± 0.91
After 3rd month	2.88 ± 0.63	4.15 ± 0.97

Table 9.
Dynamics of pain during stair climbing (Mean ± SD).

In the CG after completion of the treatment course, pain during ascending stairs decreased ($p < 0.001$). The therapeutic effect persisted until the end of the first month, after which there was an increase in pain and a statistical difference between completion of therapy and month 3, and between months 1 and 3 ($p < 0.001$).

After analyzing the results obtained, it was found that the complex rehabilitation carried out in the Therapeutic group had a longer lasting effect, with retention of the pain level at month 3 at the level immediately after therapy (**Table 9**).

3.5 Comparative analysis Western Ontario and McMaster universities osteoarthritis index (WOMAC index)

The total WOMAC Index (Mean ± SD (range)) at baseline for Therapeutic Group are (61.5 ± 8.2 (40–78)), for Control Group (59.0 ± 7.2(46–77)). No statistical difference was observed before therapy for the total WOMAC Index between groups ($p = 0.125$) (**Table 10**).

In the 3rd month after therapy, there was a statistically significant improvement in total WOMAC Index for TG patients (46 ± 4.5(36.0–54.0)) compared to CG

WOMAC index (Me(Range))	Therapeutic group	Control group	Statistical difference
WOMAC Index	62(40–78)	59(46–77)	p = 0.125
WOMAC Pain	13(7.0–16)	13(7.0–15.0)	p = 0.684
WOMAC Stiffness	5(2.0–7.0)	5(3.0–7.0)	p = 0.098
WOMAC Function	45(36.0–56.0)	43(34.0–56.0)	p = 0.015

Table 10.
Total WOMAC index and subscales pain, stiffness, and function (Me(Range)) before the therapy, divided into groups, at baseline.

WOMAC index (Me(Range))	Therapeutic group	Control group	Statistical difference
WOMAC Index	47(36–54)	56(46–73)	p < 0.001
WOMAC Pain	8(5–10)	11(8–14)	p < 0.001
WOMAC Stiffness	3(1–4)	4(3–6)	p < 0.001
WOMAC Function	35(28–42)	42(34–56)	p < 0.001

Table 11.
Total WOMAC index and subscales pain, stiffness, and function (Me (range)) 3 months after therapy, and dynamics compared to the beginning of therapy, divided into groups.

patients (56.2 ± 7.0(46.0–73.0)) and a statistically significant difference between the groups ($p < 0.001$) in favor of the group with complex rehabilitation with DO and KT (**Table 11**).

4. Discussion and findings

The conducted research project aimed to investigate the therapeutic effects of the low-frequency and low intensity pulsed electrostatic field application using the DO-therapy in patients with knee osteoarthritis and to compare the therapeutic effects of the application of Deep Oscillation® and placebo-DO and kinesitherapy (KT). The relevant value of the reformed physical factor on the basis of the KT was evaluated.

A statistically significant decrease in edema, increased range of motion (flexion) in knee joint, and m. Quadriceps femoris muscle weakness for three-time periods (TTP) versus baseline was observed in the Therapeutic Group, as well as a tendency to a better influence of the levels of muscle weakness and range of motion in the KJ for the Therapeutic group in which patients received complex physiotherapeutic treatment: electro-procedure and KT.

A statistically significant reduction in pain after treatment in Therapeutic Group and 1st and 3rd months at rest, motion, descent, and ascending were reported compared to baseline values and retention of posttreatment outcomes and at the 3rd month.

The Control Group showed a reduction in pain at rest for the three time periods (TTP), compared to baseline values, increase in the 3rd month compared to post-treatment, pain-motion reduction after treatment, and increase in the 3rd month compared to post-treatment, pain when descending stairs reduction for TTP, compared to baseline, increase in 3rd month versus post-therapy, pain-upstairs decrease in TTP, and retention of the effect achieved up to the 1st month after treatment.

The total WOMAC Index for Therapeutic Group shows a statistically significant reduction in the TTP score as well as retention of the effect on the 3rd month post-therapy versus post treatment was observed. In Control Group, a significant reduction after treatment on 1st month was observed and no statistically significant difference on 3rd month compared to baseline was found.

The frequency used in Deep Oscillation® determines the therapeutic effects achieved by the application of a low-frequency and low-intensity electrostatic field. The reduction of pain is an effect that appears relatively quickly and has a long-lasting effect [5, 6], which is probably due to the inclusion of the mechanism for Gate control of pain nociception [29], and a psychological effect of the conducted therapeutic intervention [69]. Oxidation processes in lipid substances are suppressed and also probably the formation of oxygen-containing reactive species, which is probably the basis for the manifestation of the anti-inflammatory effect [70, 71].

Microcirculation in the affected region was improved by reducing inefficient shunt blood flow and dilating capillary blood vessels [72, 73]. The observed long-term vasodilatation is a consequence of the biologically active substances (histamine, protease, serotonin, and heparin) released by the mast cells [73]. A reduction in the swelling of the tissues is found as a result of improved lymphatic drainage [2, 70]. The tissues increase their elasticity and the possibility of developing fibrosis is reduced, and the range of motion in the joints is also improved. In the area of application of low-frequency and low-intensity electrostatic field, muscle spasm is reduced, probably through a mechanism of impact on the nerve endings and the nerve conductors. The application of Deep Oscillation® leads to improvement of trophic and restorative processes in the tissues, by activating local blood flow, improving local microhemodynamics, optimizing blood transport functions, and improving the exchange that takes place through the capillaries [4, 7].

The therapeutic results achieved in placebo treatment have been associated by a number of authors with various etiological factors, such as gender, age, diagnosis, study design, trust in the therapist, and sociocultural characteristics of the ethnicity [74, 75]. One mechanism of the placebo effect has been associated with an expectancy of an outcome from the therapy conducted [76]. Expectancy associated with the treatment effect represents the likelihood that the procedure or drug substance will have an effect in relieving pain [77]. If the patient has an expectation of pain reduction after administration of a placebo treatment, this fact alone may lead to pain relief [78, 79]. Scientific evidence suggests an association with the endogenous opioid system for the expression of the placebo effect in the setting of an expectation of pain reduction [80, 81]. For this reason, an effective treatment should be considered to be one in which the analgesic effect is greater than that of placebo therapy [82].

A number of authors have reported that patients with KJ osteoarthritis suffer from Quadriceps muscle weakness, proprioception deficits, loss of balance, and consequently are at risk of frequent falls. Muscle weakness is often associated with the level of pain and functional limitations. It is assumed that these disorders are caused by a decrease in voluntary contractions with arthrogenic genesis and reflex inhibition as a result of pain and edema [83, 84]. These symptoms of osteoarthritis, especially of the knee, can be reduced or prevented by therapeutic exercises, for which there is significant evidence. Age-related changes in the neuromuscular system affect motor function in adults over 60 and especially over 80 years. Reduced muscle strength, slower contractile speed, increased fatigue, and reduced joint stability, which can vary in different individuals, are observed [85]. The extent to which muscle weakness and atrophy are caused by KJ degeneration or muscle weakness precedes it is discussed [55]. Quadriceps femoris muscle weakness is currently thought to be a predictor of OA more often in women [86], and the role of afferent sensory dysfunction is important for the progression of KOA. Muscle function is more related to joint pain than narrowing of the joint space and is easier to change, making it a realistic therapeutic goal [55, 87, 88]. Physical activity can change the properties and function of the motor unit in adults, although the effects on the variability of motor characteristics are largely unknown [85].

There is considerable evidence for the effectiveness of therapeutic exercises in knee joint osteoarthritis. A summary of a large number of systematic studies evaluates the effect of exercise and identifies improvements in pain, function, and overall assessment [89]. Therapeutic exercises are likely to prevent degenerative changes in cartilage, to reduce inflammatory activity and changes in the subchondral and metaphyseal areas of bone. There is growing evidence that therapeutic exercises can affect pain, stiffness, muscle weakness, and joint dysfunction in KOA. Therapeutic options include exercises to increase muscle strength, aerobic exercises, neuromuscular exercises, balance exercises, proprioception, water exercises, and some traditional exercises [90].

Induced endogenous analgesia as a result of therapeutic exercise is thought to be due to the release of endogenous opioids and growth factors [91], as well as the activation of cerebrospinal nociceptive inhibitory mechanisms controlled by the brain [92]. The improvement can also be explained with the biomechanical changes in the joint and increased stability [93]. The application of exercises to increase muscle strength (with/without other types of therapeutic exercises), depending on their characteristics (type of resistance, type of muscle contraction, intensity, and duration) can significantly reduce pain and improve physical function and quality of life in knee osteoarthritis. Combining them with other therapeutic agents, such as patellar taping or manual therapy, is subject to further study. It is necessary to develop a combination therapy including behavioral strategy and exercises to increase muscle strength, which has a longer-lasting effect [94]. Positive results from the combined application of Deep Oscillation and therapeutic exercises have also been reported by other research groups in recent years [95–98].

The conducted study confirms the results obtained in the complex rehabilitation of patients with knee arthritis involving the application of Deep Oscillation® therapy. Some authors report reported reduction in pain and improved movement in patients with osteoarthritis in a complex with targeted kinesitherapy [35, 38, 95]. Pain-related decline in physical function further exacerbates permanent disability in patients with knee osteoarthritis. Most patients require oral NSAIDs for pain relief, which increases the risk of adverse effects: gastrointestinal, cardiovascular, or renal. Non-pharmacological treatment modalities are recommended: topical NSAIDs, physiotherapy, therapeutic exercise, weight reduction, orthoses, intraarticular administration of corticosteroids, hyaluronic acid, and platelet-rich plasma PRP therapy [54].

The long-lasting effect we found in reducing pain at rest and during physical activity, improving range of motion and reducing knee joint swelling for the Therapeutic Group, is probably due to the complex use of low-frequency and low-intensity pulsed electrostatic field (Deep Oscillation®) and the applied therapeutic exercise program. The effect achieved in the Control group is less pronounced and is probably based on the effect of the applied kinesitherapy and the psychological effect shown by the application of the placebo-Deep Oscillation®.

5. Conclusion

Deep Oscillation® effectively reduces edema, stiffness, pain in rest position, and physical activity in knee osteoarthritis. The complex application of low-frequency and low-intensive electrostatic field with therapeutic exercise permanently affects the muscle imbalance and the range of motion in the affected joint. The established therapeutic effects of Deep Oscillation® have a long-term effect (at least 3 months) versus placebo-DO. The use of Deep Oscillation® in rehabilitation programs leads to improvement of the psycho-emotional state, functional activity, and quality of life (WOMAC Index).

There was no established discomfort in the patients treated with Deep Oscillation. Compliance with safety rules reduces the risk of side effects. The proven methodology for therapy has an easy practical application. For better objectifying of the effects of the low-frequency and low-intensity electrostatic field, the studies should continue.

Author details

Galina Mratskova
Department of Kinesitherapy and Rehabilitation, Physical Medicine and Sports, Faculty of Medicine, Trakia University, Stara Zagora, Bulgaria

*Address all correspondence to: galina.mratskova@trakia-uni.bg

References

[1] Atkinson R. A simple theory of the Johnsen-Rahbek effect. Journal of Physics D: Applied Physics. 1969;**2**(3):325. DOI: 10.1088/0022-3727/2/3/303

[2] Jahr S, Schoppe B, Reisshauer A. Effect of treatment with low-intensity and extremely low-frequency electrostatic fields (Deep Oscillation) on breast tissue and pain in patients with secondary breast lymphoedema. Journal of Rehabilitation Medicine. 2008;**40**(8):645-650. DOI: 10.2340/16501977-0225

[3] Razumov AN, Pogonchenkova IV, Khan MA, Lyan NA, Vakhova EL, Mikitchenko NA. The application of the pulsed low-frequency electrostatic field in pediatrics. Voprosy Kurortologii, Fizioterapii, i Lechebnoĭ Fizicheskoĭ Kultury. 2019;**96**(1):55-62. DOI: 10.17116/kurort20199601155

[4] Locheva V, Todorov I, Panayotova-Ovcharova L. Therapy with Deep Oscillations—Principle, biological effects. Review. Varna Medical Forum. 2019;**8**(2):105-114. DOI: 10.14748/vmf.v8i2.5974

[5] Fistetto G, Iannitti T, Capone S, Torricelli F, Palmieri B. Deep Oscillation®: Esperienze terapeutico-riabilitative con un nuovo innovativo strumento ad azione elettrostatica [Deep Oscillation: Therapeutic-rehabilitative experiences with a new electrostatic device]. Minerva Medica. 2011;**102**(4):277-288

[6] Kraft K, Kanter S, Janik H. Safety and effectiveness of vibration massage by Deep Oscillations: A prospective observational study. Evidence-Based Complementary and Alternative Medicine. 2013;**2013**:679248. DOI: 10.1155/2013/679248

[7] Kulikov AG, Yarustovskaya OV, Kuzovleva EV, Zaitseva TN, Kulchitskaya DB, Konchugova TV. Application of low-frequency electrostatic field in clinical practice. Russian Journal of Physiotherapy, Balneology and Rehabilitation. 2019;**8**(3):195-209. DOI: 10.17816/1681-3456-2019-18-3-195-209

[8] Aliyev R, Mikus EWS, Reinhold JG. Highly significant therapeutic success with Deep Oscillation® in orthopedic rehabilitation. Orthopedic Practice. 2008;**44**:448-453

[9] Aliyev R. Klinische Wirksamkeit des Therapieverfahrens Tiefenoszillation bei Sportverletzungen [clinical effects of the therapy method Deep Oscillation in treatment of sports injuries]. Sportverletzung Sportschaden. 2009;**23**(1):31-34. DOI: 10.1055/s-0028-1109216

[10] Mratskova G. Comparative study of the Deep Oscillation therapy effects and some preformed physical factors in treatment and rehabilitation of patients with osteoarthritis [thesis]. Stara Zagora: Trakia University, Faculty of Medicine; 2018

[11] Mratskova G. Use of Deep Oscillation- therapy in rehabilitation program for patient after distal radius fracture with a complex regional pain syndrome (sudeck atrophy): A case report. Trakia Journal of Sciences. 2020;**18**(1):187-193. DOI: 10.15547/tjs.2020.s.01.034

[12] Reinhold J, Deeva I, Korkina L, Schaper K, Krummenauer F. Randomisierte Pilotstudie zur Quantifizierung des patientenseitigen Nutzens der Beeinflussung primärer

Wundheilungsprozesse durch Tiefenoszillation [Randomised pilot study for quantification of benefit from the patient's point of view of Deep Oscillation treatment in primary wound healing]. Zeitschrift für Orthopädie und Unfallchirurgie. 2014;**152**(3):260-264. DOI: 10.1055/s-0034-1368447

[13] O'Brien CP, Watson A. Deep Oscillation® therapy in the treatment of lateral epicondylalgia: A pilot randomized control trial. Journal of Sports Medicine and Doping Studies. 2016;**6**(3):180. DOI: 10.4172/2161-0673.1000180

[14] Mratskova G, Dimitrov PD. Effectiveness of complex rehabilitation with Deep Oscillation and kinesitherapy for pain relief in patients with gonarthritis. KNOWLEDGE—International Journal. 2018;**26**(4):1071-1078. Available from: http://ikm.mk/ojs/index.php/kij/article/view/3086

[15] Vladeva E, Mihaylova M, Panayotova L. Deep Oscillations-reducing edema and improving kinesiology in the early stages after knee joint arthroplasty. Journal of IMAB. 2021;**27**(1):3577-3581. DOI: 10.5272/jimab.2021271.3577

[16] Koleva IB, Yoshinov BR, Asenova TA, Yoshinov RR. Physical analgesia: Methods, mechanisms and algorithms for post-operative Pain [Internet]. In: Topics in Postoperative Pain. IntechOpen; 2023. DOI: 10.5772/intechopen.111590

[17] Hausmann M, Ober J, Lepley AS. The effectiveness of Deep Oscillation therapy on reducing swelling and pain in athletes with acute lateral ankle sprains. Journal of Sport Rehabilitation. 2019;**28**(8):902-905. DOI: 10.1123/jsr.2018-0152

[18] Boisnic S, Branchet MC. Anti-inflammatory and draining effect of the Deep Oscillation® device tested clinically and on a model of human skin maintained in survival condition. European Journal of Dermatology. 2013;**23**(1):59-63. DOI: 10.1684/ejd.2012.1904

[19] Khodyreva LA, Dudareva AA, Kumachev KV, et al. New physical factors in the treatment of chronic abacterial prostatitis/chronic pelvic pain syndrome. Handbook for the General Practitioner. 2012;**3**:23-28

[20] Teo I, Coulborn A, Munnoch DA. Use of the HIVAMAT® 200 with manual lymphatic drainage in the management of lower limb lymphoedema and lipoedema. Journal of Lymphoedema. 2016;**11**(1):49-53

[21] Kashilska Y, Petkov A, Micheva P, Batashki A, Batashk Z. Improving the quality of life through effects of treatment with low-intensity extremely low-frequency electrostatic field with DEEP OSCILLATION® in patients breast cancer with secondary limb to patients treated with standard lymphatic equipment. Medicine. 2015;**1**:381-387. Available from: https://api.semanticscholar.org/CorpusID:36465294

[22] Belmonte R, Tejero M, Ferrer M, et al. Efficacy of low-frequency low-intensity electrotherapy in the treatment of breast cancer-related lymphoedema: A cross-over randomized trial. Clinical Rehabilitation. 2012;**26**(7):607-618. DOI: 10.1177/0269215511427414

[23] De Leon PH. A systematic review of Deep tissue Oscillation and a study proposal of its effects on geriatric bedridden patients with lower extremity lymphedema. [thesis] Texas State University. 2016

[24] Gao Y-C, Peng CC, Peng RY. A long term chronic fibrotic adhesion of

elbow muscles alleviated by applying hivamate 200 Deep Oscillation® therapy. International Journal of Multidisciplinary Research and Development. 2015;**2**(1):286-289

[25] Tápanes SH, Fernández MJS, Iturralde Y, Fernández AS. The effect of Deep Oscillation therapy in fibrocystic breast disease. A randomized controlled clinical trial. International Archives of Medicine. 2018;**11**(14):1-10. DOI: 10.3823/2555

[26] Hinman MR, Lundy R, Perry E, Robbins K, Quarter L. Comparative effect of ultrasound and DEEP OSCILLATION® on the extensibility of hamstring muscles. Journal of Athletics Medicine. 2013;**1**(1):45-55

[27] Winkelmann ZK, Roberts EJ, Games KE. Acute effects and perceptions of Deep Oscillation therapy for improving hamstring flexibility [published correction appears in Journal of Sport Rehabilitation. 2019 May 1;28(4):403]. Journal of Sport Rehabilitation. 2018;**27**(6):570-576. DOI: 10.1123/jsr.2017-0044

[28] Mratskova G, Petrov D, Dimitrov N. Short term effects of low- frequency and low intensity electrostatic field in patients with knee joint osteoarthritis. KNOWLEDGE—International Journal. 2018;**28**(2):547-551. DOI: 10.35120/kij2802547M

[29] Trybulsky R. Using Deep Oscillation® system in the treatment of wounds. Rehabilitacja w Praktyce. 2008;**1**:28-33

[30] Reinhold J. Mechanisms of Deep Oscillation®. Manual Lymphatic Drainage UK. UK [WWW document.]. 2017. Available from: http://www.physiopod.co.uk/assets/images2017/2017-MLD-6pp-web.pdf

[31] Afanasyeva TN. Variable low-frequency electrostatic field in the complex treatment of patients with pneumonia [Abstract of PhD thesis]. Moscow; 2004

[32] Yashkov AV, Gazdieva EM, Badyanova IS. Efficacy of intermittent low-frequency electrostatic field in the sanatorium-based complex treatment of patients with chronic obstructive pulmonary disease. Kurortniye Vedmosti. 2007;**3**(42):62-63

[33] Zehtindjieva MG, Ioshinov BR, Andonov DR, Ilkov VS, Bayraktarova A, Koleva IB. Deep Oscillation—A modern additional physical modality for analgesia in patients with back pain. Praemedicus. 2013;**29**:85-90

[34] Kulchitskaja DB, Orehova EM, Konchugova TV, Kolbaja LI. Application of the impulse electric field for the treatment of patients with gonarthrosis. Physiotherapist. 2014;**2**:53-57

[35] Coronados VY, Alba GCJ, Lorenzo GE, et al. Therapy of Deep Oscillations as a complement to kinesiotherapy in the treatment of gonarthrosis. Revista Cubana de Medicina Física y Rehabilitación. 2017;**9**(2):1-14

[36] Mratskova G, Dimitrov N, Goycheva P, Petrov D. Assesment of the rehabilitation program for patients with knee osteoarthritis and diabetes mellitus type 2. Knowledge International Journal. 2021;**45**(4):853-859. Available from: http://ikm.mk/ojs/i

[37] Koleva IB, Yoshinov B. Rehabilitation as an essential element in the clinical practice of orthopaedics and traumatology. Acta Scientific Orthopaedics. 2020;**3**(9):44-46

[38] Kulchitskaya DB, Konchugova TV, Salamadina MO. Non-drug methods of treatment of patients with gonarthrosis.

Journal Bulletin of Rehabilitation Medicine. 2019;**2**:40-43

[39] Vakhova EL, Vybornov DY, Turova EA, et al. Low-frequency electrostatic field application for injuries of knee joint capsular ligament apparatus in children: A prospective randomized comparative study. Bulletin of Rehabilitation Medicine. 2022;**21**(4):71-78

[40] Li AA, Korchazhkina NB, Kazantsev AB, Li EA. Influence of pulsed low-frequency electrostatic field on the activity of the inflammatory process in patients operated for hallux valgus of the first toe. Russian Journal of the Physical Therapy, Balneotherapy and Rehabilitation. 2009;**5**:60-61

[41] Vladeva E, Mihaylova M, Bacheva D, Todorov I. Our experience with using Deep Oscillation in early rehabilitation of patients with arthoplasty of the knee joint Varna medical. Forum. 2018;**7**(1):66-70. DOI: 10.14748/vmf.v7i1.4925

[42] Kulchitskaya DB, Fesyun AD, Samoilov AS, Kolbakhova SN, Gushchina NV, Astakhova KA. The use of rehabilitation programs for patients in the early recovery period after total knee arthroplasty. Problems of Balneology, Physiotherapy, and Exercise Therapy. 2022;**99**(2):32-36. (In Russ.). DOI: 10.17116/kurort20229902132

[43] Koleva IB, Ioshinov BR, Yoshinov RD. Complex analgesia (infiltrations and Deep Oscillation) in patients with stump pain and phantom pain after lower limb amputation (double-blind randomised controlled trial of efficacy). Journal of Advances in Medicine and Medical Research. 2017;**22**(11):1-17. DOI: 10.9734/JAMMR/2017/34198

[44] Cherevashchenko L, Velikanov D, Tsogoev A, Serebryakov A, Bobrik Y, Gaidamaka I, et al. Efficiency of using physical therapeutic factors in outpatient medical rehabilitation of patients with dyscirculatory encephalopathy. Azerbaijan Medical Journal. 2022;**4**:177-185. DOI: 10.34921/amj.2021.4.025

[45] Koleva IB, Yoshinov BR, Yoshinov RR. Neurorehabilitation algorithms in Parkinsonism: Impact of electrical stimulations and Deep Oscillation on autonomy and quality of life. Journal of Alzheimers & Neurodegenerative Diseases. 2020;**6**:043. DOI: 10.24966/AND-9608/100043

[46] Hunter DJ, Bierma-Zeinstra S. Osteoarthritis. Lancet. 2019;**393**(10182):1745-1759. DOI: 10.1016/S0140-6736(19)30417-9

[47] Kloppenburg M, Berenbaum F. Osteoarthritis year in review 2019: Epidemiology and therapy. Osteoarthritis and Cartilage. 2020;**28**(3):242-248. DOI: 10.1016/j.joca.2020.01.002.2

[48] Vos T, Flaxman AD, Naghavi M, et al. Years lived with disability (YLDs) for 1160 sequelae of 289 diseases and injuries 1990-2010: A systematic analysis for the global burden of disease study 2010. Lancet. 2013;**381**(9867):628

[49] GBD 2013 DALYs and HALE Collaborators, Murray CJ, Barber RM, et al. Global, regional, and national disability-adjusted life years (DALYs) for 306 diseases and injuries and healthy life expectancy (HALE) for 188 countries, 1990-2013: Quantifying the epidemiological transition. Lancet. 2015;**386**(10009):2145-2191. DOI: 10.1016/S0140-6736(15)61340-X

[50] Lawrence RC, Felson DT, Helmick CG, et al. Estimates of the prevalence of arthritis and other rheumatic conditions in the United

States. Part II. Arthritis Rheum. 2008;**58**(1):26-35. DOI: 10.1002/art.23176

[51] Wallace IJ, Worthington S, Felson DT, et al. Knee osteoarthritis has doubled in prevalence since the mid-20th century. Proceedings of the National Academy of Sciences of the United States of America. 2017;**114**(35):9332-9336. DOI: 10.1073/pnas.1703856114

[52] Loeser RF. Age-related changes in the musculoskeletal system and the development of osteoarthritis. Clinics in Geriatric Medicine. 2010;**26**(3):371-386. DOI: 10.1016/j.cger.2010.03.002

[53] Hsu H, Siwiec RM. Knee osteoarthritis. In: StatPearls. Treasure Island (FL): StatPearls Publishing; 2021. Last Update: September 4, 2022

[54] Huang KH, Hsieh RL, Lee WC. Pain, physical function, and health in patients with knee osteoarthritis. Rehabilitation Nurse. 2017;**42**(4):235-241. DOI: 10.1002/rnj.234

[55] Roos EM, Herzog W, Block JA, Bennell KL. Muscle weakness, afferent sensory dysfunction and exercise in knee osteoarthritis. Nature Reviews Rheumatology. 2011;**7**(1):57-63. DOI: 10.1038/nrrheum.2010.195

[56] Lawrence JS, Bremner JM, Bier F. Osteo-arthrosis prevalence in the population and relationship between symptoms and x-ray changes. Annals of the Rheumatic Diseases. 1966;**25**:1-24

[57] Slemenda C, Brandt KD, Heilman DK, et al. Quadriceps weakness and osteoarthritis of the knee. Annals of Internal Medicine. 1997;**127**(2):97-104. DOI: 10.7326/0003-4819-127-2-199707150-00001

[58] Emery CA, Whittaker JL, Mahmoudian A, et al. Establishing outcome measures in early knee osteoarthritis. Nature Reviews Rheumatology. 2019;**15**(7):438-448. DOI: 10.1038/s41584-019-0237-3

[59] Woolf AD, Pfleger B. Burden of major musculoskeletal conditions. Bulletin of the World Health Organization Journal. 2003;**81**(9):646-656

[60] Sebbag E, Felten R, Sagez F, Sibilia J, Devilliers H, Arnaud L. The world-wide burden of musculoskeletal diseases: A systematic analysis of the World Health Organization burden of diseases database. Annals of the Rheumatic Diseases. 2019;**78**(6):844-848. DOI: 10.1136/annrheumdis-2019-215142

[61] Nguyen US, Zhang Y, Zhu Y, Niu J, Zhang B, Felson DT. Increasing prevalence of knee pain and symptomatic knee osteoarthritis: Survey and cohort data. Annals of Internal Medicine. 2011;**155**(11):725-732. DOI: 10.7326/0003-4819-155-11-201112060-00004

[62] Felson DT. Osteoarthritis as a disease of mechanics. Osteoarthritis and Cartilage. 2013;**21**(1):10-15. DOI: 10.1016/j.joca.2012.09.012

[63] Robinson WH, Lepus CM, Wang Q, et al. Low-grade inflammation as a key mediator of the pathogenesis of osteoarthritis. Nature Reviews Rheumatology. 2016;**12**(10):580-592. DOI: 10.1038/nrrheum.2016.136

[64] Palazzo C, Nguyen C, Lefevre-Colau MM, Rannou F, Poiraudeau S. Risk factors and burden of osteoarthritis. Annals of Physical Medicine and Rehabilitation. 2016;**59**(3):134-138. DOI: 10.1016/j.rehab.2016.01.006

[65] Shorter E, Sannicandro AJ, Poulet B, Goljanek-Whysall K. Skeletal

muscle wasting and its relationship with osteoarthritis: A mini-review of mechanisms and current interventions. Current Rheumatology Reports. 2019;**21**(8):40. DOI: 10.1007/s11926-019-0839-4

[66] Ilieva E, Minchev R. The role of patient education in complex treatment and rehabilitation in patients with degenerative joint diseases. Physical Medicine, Rehabilitation, Health. 2010;**2**:17-226

[67] Cross M, Smith E, Hoy D, et al. The global burden of hip and knee osteoarthritis: Estimates from the global burden of disease 2010 study. Annals of the Rheumatic Diseases. 2014;**73**:1323-1330

[68] Frontera WR, DeLisa JA, Gans BM, Robinson LR, Bockenek W, Chae J. DeLisa's Physical Medicine and Rehabilitation: Principles and Practice. Wolters Kluwer Health/Lippincott Williams & Wilkins. ISBN/ISSN 9781496374967. Publication Date July 9, 2019

[69] Koleva I. Basics of Physical Medicine, Physical Therapy and Rehabilitation. Steno, Sofia, Medical University of Pleven; 2013. ISBN: 978-954-756-103-8

[70] Gasbarro V et al. Role of Hivamat® (Deep Oscillation) in the treatment for the lymphedema of the limbs. European Journal of Lymphology. 2006;**16**:13-18

[71] Mikhalchik E et al. Effects on bloodparameters of DEEP OSCILLATION®. In: 1st International Conference on Skin and Environment. Vol. 59. Moscow: St.Petersburg; 2005

[72] Korkina L, Reinhold J, Rota L, Primavera G, Raskovic D. Treatment of gynoid lipodystrophy (Cellulite) with DEEP OSCILLATION®: A pilot clinical study. In: 29th Annual Meeting of The Bioelectromagnetics Society 29th Annual Meeting Abstract Collection; 10-15 June 2007; Kanazawa, Japan. 2007. pp. 30-31. Available from: https://www.yumpu.com/en/document/read/6108446/the-bioelectromagnetics-society-29-annual-meeting-abstract-

[73] Starosvetskaya OA. The application of a pulsed low-frequency electrostatic field in the patients presenting with hypertensive type neurocirculatory dystonia. Russian Journal of Physiotherapy, Balneology and Rehabilitation. 2013;**12**(6):39-46. DOI: 10.17816/41430

[74] Johnson MI, Ashton CH, Bousfield DR, Thompson JW. Analgesic effects of different frequencies of transcutaneous electrical nerve stimulation on cold-induced pain in normal subjects. Pain. 1989;**39**(2):231-236. DOI: 10.1016/0304-3959(89)90010-9

[75] Moffett J, McLean S. The role of physiotherapy in the management of non-specific back pain and neck pain. Rheumatology (Oxford). 2006;**45**(4):371-378. DOI: 10.1093/rheumatology/kei242

[76] Pollo A, Amanzio M, Arslanian A, Casadio C, Maggi G, Benedetti F. Response expectancies in placebo analgesia and their clinical relevance. Pain. 2001;**93**(1):77-84. DOI: 10.1016/S0304-3959(01)00296-2

[77] Vase L, Robinson ME, Verne GN, Price DD. The contributions of suggestion, desire, and expectation to placebo effects in irritable bowel syndrome patients. An empirical investigation. Pain. 2003;**105**(1-2):17-25. DOI: 10.1016/s0304-3959(03)00073-3

[78] Požgain I, Požgain Z, Degmečić D. Placebo and nocebo effect: A mini-review. Psychiatria Danubina. 2014;**26**(2):100-107

[79] Annoni M. Better than nothing: A historical account of placebos and placebo effects from modern to contemporary medicine. International Review of Neurobiology. 2020;**153**:3-26. DOI: 10.1016/bs.irn.2020.03.028

[80] Scott DJ, Stohler CS, Egnatuk CM, Wang H, Koeppe RA, Zubieta JK. Placebo and nocebo effects are defined by opposite opioid and dopaminergic responses. Archives of General Psychiatry. 2008;**65**(2):220-231. DOI: 10.1001/archgenpsychiatry.2007.34

[81] Büchel C. The role of expectations, control and reward in the development of pain persistence based on a unified model. eLife. 2023;**12**:e81795. Published 2023 Mar 27. DOI: 10.7554/eLife.81795

[82] Wigley R. When is a placebo effect not an effect? Clinical Medicine (London). 2007;7(5):450-452. DOI: 10.7861/clinmedicine.7-5-450

[83] Roddy E, Doherty M. Changing life-styles and osteoarthritis: What is the evidence? Best Practice & Research: Clinical Rheumatology. 2006;**20**(1):81-97

[84] Guan B, Liu F, Mizaian AH, et al. Deep learning approach to predict pain progression in knee osteoarthritis. Skeletal Radiology. 2022;**51**(2):363-373. DOI: 10.1007/s00256-021-03773-0

[85] Hunter SK, Pereira HM, Keenan KG. The aging neuromuscular system and motor performance. Journal of Applied Physiology 1985. 2016;**121**(4):982-995. DOI: 10.1152/japplphysiol.00475.2016

[86] Segal NA, Glass NA, Felson DT, et al. Effect of quadriceps strength and proprioception on risk for knee osteoarthritis. Medicine & Science in Sports & Exercise. 2010;**42**(11):2081-2088. DOI: 10.1249/MSS.0b013e3181dd902e

[87] Øiestad BE, Juhl CB, Eitzen I, Thorlund JB. Knee extensor muscle weakness is a risk factor for development of knee osteoarthritis. A systematic review and meta-analysis. Osteoarthritis and Cartilage. 2015;**23**(2):171-177. DOI: 10.1016/j.joca.2014.10.008

[88] Roos EM, Arden NK. Strategies for the prevention of knee osteoarthritis. Nature Reviews Rheumatology. 2016;**12**(2):92-101. DOI: 10.1038/nrrheum.2015.135

[89] Ilieva EM, Oral A, Küçükdeveci AA, et al. Osteoarthritis. The role of physical and rehabilitation medicine physicians. The European perspective based on the best evidence. A paper by the UEMS-PRM section professional practice committee. European Journal of Physical and Rehabilitation Medicine. 2013;**49**(4):579-593

[90] Zeng CY, Zhang ZR, Tang ZM, Hua FZ. Benefits and mechanisms of exercise training for knee osteoarthritis. Frontiers in Physiology. 2021;**12**:794062. DOI: 10.3389/fphys.2021.794062

[91] Koltyn KF. Exercise-induced hypoalgesia and intensity of exercise. Sports Medicine. 2002;**32**(8):477-487. DOI: 10.2165/00007256-200232080-00001

[92] Ray CA, Carter JR. Central modulation of exercise-induced muscle pain in humans. Journal of Physiology. 2007;**585**(Pt 1):287-294. DOI: 10.1113/jphysiol.2007.140509

[93] King LK, Birmingham TB, Kean CO, Jones IC, Bryant DM,

Giffin JR. Resistance training for medial compartment knee osteoarthritis and malalignment. Medicine & Science in Sports & Exercise. 2008;**40**(8):1376-1384. DOI: 10.1249/MSS.0b013e31816f1c4a

[94] Brosseau L, Taki J, Desjardins B, et al. The Ottawa panel clinical practice guidelines for the management of knee osteoarthritis. Part one: Introduction, and mind-body exercise programs. Clinical Rehabilitation. 2017;**31**(5):582-595

[95] Valeva S, Sazdova L. Role of elastic resistance, manual mobilization techniques and Deep Oscillation in the treatment of ankle fractures. KNOWLEDGE—International Journal. 2023;**61**(4):683-687. Available from: https://ikm.mk/ojs/index.php/kij/article/view/6474

[96] Whizar-Lugo M, Domínguez-Franco V, Minutti-Palacios A, Dominguez-Cherit G. editors. Topics in Postoperative Pain [Internet]. London, UK: IntechOpen; 2023. DOI: 10.5772/intechopen.105661

[97] Villalba-Meneses F, Chaglla-Monge K, Almeida-Galárraga D, et al. Evaluation of Deep Oscillation therapy for the treatment of lumbar pain syndrome using motion capture systems: A systematic review. Journal of Bodywork & Movement Therapies. 2024;**38**:180-190. DOI: 10.1016/j.jbmt.2024.01.010

[98] Onose G et al. reported reduction in pain and improved movement in patients with osteoarthritis in a complex with targeted kinesitherapy In: Onose GC, Daia-Chendreanu et al, editors. Our experience concerning the use of Deep Oscillation in the therapy of the use of Deep Oscillation in the therapy of neurologic patients with associated pain and/or rheumatic—degenerative conditions. The proceedings from the National Conference of Neurosurgery and Neurorehabilitation, with international participation, 2nd edition, Mamaia, Romania, 9-12 septembrie, 2009. pp 173-176; supliment al "Industria Textile Magazine" cotat ISI, Ed. Certex, 2009

Chapter 2

Experimental Characterization and Data Analysis of Electrostatic Discharge (ESD) with Microenvironment Dominated by Electrode Approaching Velocity Based on Wavelet Neural Network

Yang Meng, Fangming Ruan, Yan Lyu, Ze Chen, Dandan Qu and Kun Zhang

Abstract

The microenvironment dominated by the electrode approaching velocity can be defined as the micro-gap ESD model. Micro-gap ESD is a time-domain continuous process, and the electrode approach velocity critically influences the characteristic discharge parameters. Based on two sub-processes, field electron emission and Townsend gas discharge, the experimental characteristic parameters are analyzed in time frequency. Wavelet neural network algorithm is also used to analyze and predict the discharge results. Remarkable variation appeared in discharge parameters of discharge current peak values, current derivative peak values, and arc length due to electrode approach velocity change in experimental measurement results of micro-gap electrostatic discharge. The higher the velocity, the larger the variation of discharge parameters. Threshold phenomenon emerged for discharge parameters variation with charge voltage due to electrode moving to the target at various velocities. For micro-gap ranging in 80–300 micrometers, electrode moving velocity enhanced the sub-process of electron surface emission. The approaching velocity is an important input index, and the wavelet neural network algorithm is used to analyze and predict the output results. The characteristics measured in the real experiment were compared with the prediction parameters from the wavelet neural network calculation result. According to the prediction data, a discussion was conducted on correctness accuracy and the discharge process trend analysis. Nonlinearity and discreteness are the main characteristics of discharge parameters.

Keywords: characteristic parameters, electrode velocity, ESD, micro-gap, nonlinearity and discreteness, parameter, wavelet neural network

1. Introduction

Electrostatic discharge (ESD), dominated by approaching velocity, is a serious threat to electronic systems. Several aspects of electrostatic discharge in micro-gap, including the influence of humidity, arc length, and moving speed, have been involved in Refs. [1–5]. However, problems with the three features below need to be understood and clarified: (i) short discharge gap in the range of μm-cm, (ii) charge voltage in 200 V–2000 V, and (iii) effect on discharge parameters of the electrode (charged body) moving at a certain velocity to the target. Previous studies have not mentioned how the electrode approach velocity affects the discharge parameters, and its physical mechanism needs to be further analyzed and clarified.

Our work concentrated on discussing electrostatic discharge (ESD) events with these three features. The discharge environment is important in electrostatic discharge, including vacuum and gas states. Previous studies have focused on these two conditions. In a strict sense, the micro-gap discharge model does not belong to the above two types but is more like a combination of the two types. The discharge process can be divided into field electron emission and Townsend gas discharge. The theoretical analysis of the model combined with the two remains to be improved, especially the influence of the approach velocity on the discharge parameters is worth exploring and studying. The charging voltage ranges from 0.1 to 3 kV. We also optimized the wavelet neural network algorithm to fit the influence of various input variables on the discharge parameters, such as electrode speed, temperature, and air pressure. The influence of electrode approaching velocity on discharge parameters is verified especially, which coincides with the above theoretical analysis and experimental test results.

Electronic systems need to consider the protection of threats from electrostatic discharge. The International Electrotechnical Commission has issued IEC61000–4-2 as the test standard of electrostatic discharge immunity for electrical and electronic products. Various factors can impact the result of electrostatic discharge. However, the test results in real processes applying IEC61000–4-2 usually have comparatively low repeatability. The main reason for low reproducibility is that IEC61000–4-2 is considered electrostatic contact discharge, but many non-conducts electrostatic discharges exist in real circumstances. There are various factors in the real discharge process, for example, gas pressure, temperature, humidity, and electrode moving velocity, which may change the ESD result.

The velocity effect of the electrode moving to the target has become a hot concern in electrostatic discharge research expertise. This is because non-contacted electrostatic discharge result parameters strongly correlate with electrode moving velocity to the target. Experiment research with the electrostatic simulator (so-called ESD gun) and human/metal mode electrostatic discharge has been conducted to discuss the relationship between electrode moving velocity and discharge parameters of peak current, peak current derivative, and arc length. Micro-gap model and Townsend's gas discharge theory were applied to analyze the possible mechanisms of the velocity effect of moving electrodes on discharge parameters.

During the experiment in human/metal mode threshold phenomenon on the velocity effect of electrode moving with charge voltage emerged distinctively. The correlative coefficient's maximum positive or negative value appears when the charge voltage reaches 800 V. An illustration was given on the possible reason threshold produces at special charge voltage.

Based on the above, the benefits of this paper include by introducing a wavelet neural network, time-frequency characteristics of electrostatic discharge signal are analyzed, and denoising methods are summarized. The rule is explored between discharge conditions as input and characteristic parameters as output. The mechanism of influence of electrode approaching velocity on discharge parameters in a non-contact electrostatic discharge microenvironment is revealed.

However, there are some limitations in this paper. Non-contact electrostatic discharge is a multi-variable function with a time change. The dominant non-contact microenvironmental factors determine the main variable of the function change. In this paper, the influence of electrode approaching velocity on discharge parameters is mainly studied, and the influence of other discharge conditions remains to be further studied. The multi-factor discretization with low repetition deserves more attention.

2. The theoretical analysis

For electronic products, electrostatic discharge protection has received increasing attention. Current test standards for discharge without detailed regulation law, the effect factors of electrostatic discharge regulation are too broad, and the applicability of the different scenarios is wide enough. Therefore, given the actual cases and the complex diversity of test conditions, we model the contactless electrostatic discharge model. Among them, the factors affecting electrostatic discharge mainly include electrode approaching speed, pressure, temperature, and other discharge characteristic parameters, including discharge current peak, rising speed (when the peak current is 10 to 90%), and falling speed (when the peak current is 90 to 30%). The contactless electrostatic discharge model mainly consists of two sub-processes: the field electron emission process and the Townsend gas discharge process. How does the approach velocity affect the discharge parameters in the microenvironment dominated by the approach velocity?

Four types of electrostatic discharge events exist for different conditions, including glow, corona, electric arc, and electric spark. Electrode velocity moving to the target alters conditions of electrostatic discharge for micro-gap. The fast velocity of the electrode results in an electric spark, whereas slow velocity may be responsible for glow discharge.

The approach velocity influences the field electron emission, which indirectly affects the whole process of micro-gap discharge, and the micro-gap range is about 100 μm-10 cm. For the electrode moving at a fast velocity to the target, electric charge particles do not have enough time to form a uniform distribution on the surface of the electrode. When the electrode moves towards the target, the electric spark causes the discharge current to rise sharply, that is, the peak value of the discharge current is high and the arc length is short.

Due to the electrode's rapid movement, the discharge gun's tip accumulates a large amount of charge in a short time, forming a local high voltage. Under normal air pressure, the gas between high-voltage electrodes is broken down and produces self-excited conduction. The gas discharge phenomenon of flash and burst sound appears, namely, spark discharge. During spark discharge, collision ionization does not occur in the whole region between electrodes but along the narrow and zigzag luminous channel, accompanied by a popping sound. Due to the gas breakdown suddenly from an insulator into a good conductor, the current surged, and the power supply was insufficient, so the voltage dropped, and the discharge was temporarily extinguished until the voltage was restored to discharge again.

On the other hand, for an electrode moving at low velocity to the target, the charge on the electrode surface obtains enough time to form a uniform distribution. In a relatively closed experimental chamber, where the gas pressure is low, neutral atoms or molecules are excited by uniformly distributed charges, and the excited particles will release energy in the form of light when they fall back to the ground state from the excited state, to display the glow gas discharge phenomenon.

The mathematical model of Townsend gas ionization avalanche discharge current is described as follows [6]:

$$i(t)_{av} = C\frac{dV}{dt} + eN_0\frac{v_e}{d}\exp\left(\int_0^t \alpha(\xi)v_e(\xi)d\xi\right). \tag{1}$$

Among them,
C is the capacitance formed between two electrodes.
V is the micro-gap voltage difference.
e is the electron charge.
N_0 is the initial number of electrons supplied by the surface of the cathode.
$v_e(t)$ is the electron drift velocity relative to time t.
d is the distance between micro-gaps.
α is the Townsend first ionization coefficient.

According to (1), the greater the approaching velocity of $v_e(t)$, the greater the discharge current, and the greater the Townsend first ionization coefficient of α.

We know field electron emission is due to external electric field electron emission of the cathode surface under the action of the electric field, cathode in part through electronic barrier emitted from the surface. According to the theory of quantum mechanics, amid a trap, electrons, even if their energy is less than the potential energy barrier, also has a certain probability through the potential barrier and running away. This is called the Tunnel effect. Under the action of a strong electric field, the shape of the potential barrier becomes narrower and lower so that the electrons can easily pass through the potential barrier and be emitted from the surface.

The mathematical model of field electron emission is described as follows [6]:

$$i(t)_{surf} = \frac{4}{9}\varepsilon_0\sqrt{\frac{2e}{m_e}}\cdot\frac{S_n V^{\frac{3}{2}}}{\left(d - v_e(t)\cdot t\right)^2}K\left[\alpha(t)\right]. \tag{2}$$

Among them,
ε_0 is the vacuum permittivity.
e is the quantity of electron charge.
m_e is the electron mass.
S_n is the plasma front cross-section area.
d is the distance between micro-gaps.
V is the micro-gap voltage difference.
$v_e(t)$ is the electron drift velocity relative to time t.
$K[\alpha(t)]$ is the plasma shielding coefficient.

Parameters	Group 1	Group 2	Group 3
Process acting	Surface	Surface/gas	Gas
Arc length(μm)	<20	20–80	80–100
Breakdown voltage (V)	<700	700–2000	>1500
Electric field strength (kV/mm)	>50–100	25–75	<8–35
Current rise time (ps)	<40	300–600	<350
Current rise waveform	No step	Step	No step
Cathode effect	X	X	
Anode effect	X		
Pressure effect		X	X

Table 1.
Static discharge parameters with different micro-gap distances [7].

According to (2), the greater the electron drift velocity, the larger the micro-gap voltage difference, and the greater the discharge current. According to (1) and (2), two sub-process physical models of micro-gap electrostatic discharge are characterized by mathematical formulas in which electron drift velocity $v_e(t)$ and micro-gap voltage difference V are the main factors affecting the discharge current. At the same time, the gas pressure P of the discharge microenvironment is another important factor.

Table 1 shows the configuration of static discharge parameters for different micro-gap distances [6], where the discharge arc lengths are less than 20*um*, 20-80*um,* and 80-100*um,* respectively. Under the condition of not considering the approach velocity (i.e., static state), the discharge effect of the two sub-processes, field electron emission and Townsend gas discharge, will also change due to the change of micro-gap distance. The first group (<20*um*) is the field electron emission process, which occupies the dominant role. The second group (20-80*um*) is a combination of two sub-processes. The Townsend gas discharge process dominates the third group (80-100*um*). The breakdown voltage, electric field strength, current rise time, current rise waveform, cathode effect, anode effect, pressure effect, and other parameters are set, as shown in **Table 1**.

3. Electrostatic discharge test system based on approaching velocity

Our electromagnetic compatibility team has researched and developed novel test equipment of electrostatic discharge parameters, which can be used to discuss electrode movement speed effect. The main body of the test system is a closed box so that we can easily control the gas filling and pressure. The simulation of approaching speed is realized by controlling the crankshaft connecting rod with a stepping motor. The discharge gun is placed on the top of the connecting rod, and the speed of the stepping motor is changed to obtain different approaching speeds so that the discharge gun moves toward the discharge target and completes the discharge process. In addition, there are also gas pressure meters, high voltage supply, humidity and temperature meters, digital oscilloscopes, vacuum pumps, cameras, wireless routers, and switches. The physical picture of the experimental test system is shown in **Figure 1**. The discharge target is installed on the side wall of the sealed box, the motor holder

Figure 1.
Physical drawing of the experimental test system.

is fixed to the case, and a discharge gun is attached to the guide rail between the motor and the discharge target [6, 7]. The closed box is also provided with a pipeline connected to the vacuum pump, and a vacuum pump can be connected to adjust the air pressure inside the closed box. The negative pressure meter displays the negative pressure value in the closed box in real time. The discharge target is connected to the digital oscilloscope through a coaxial cable, and the model is Tektronix MSO56 so that

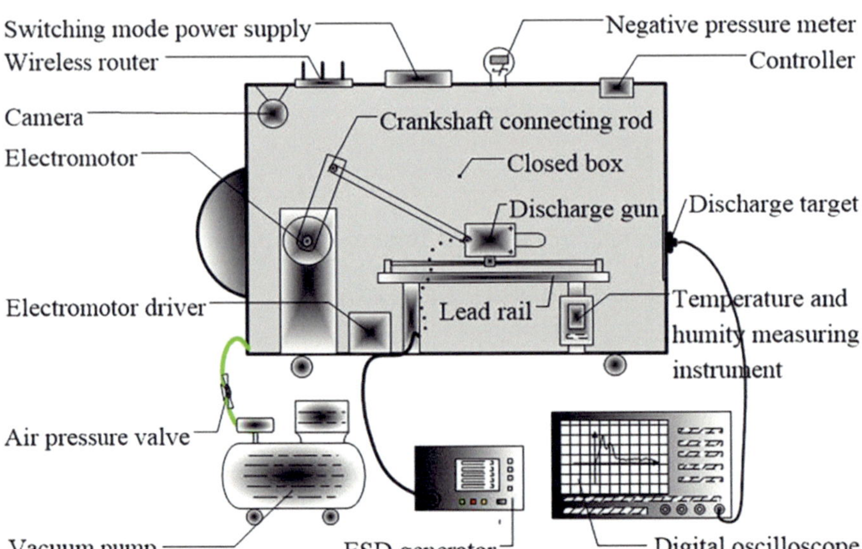

Figure 2.
Schematic diagram of structure and function of electrode approach velocity effect electrostatic discharge test system.

we can collect the discharge parameters and analyze the data. The structure and function diagram of the whole test system is shown in **Figure 2**, including the instrument configuration inside and outside the box.

The materials involved in this paper include the electrode metal of the discharge gun tip, discharge target metal, and environmental gas, among which the metal material is mainly nickel-chromium alloy, and the environmental gas is mainly air. As metal materials are exposed to the air for a long time, an oxide film has been formed on their surface, but the metal oxide film has little effect on ESD. Metal and its oxide film are the main carriers of static charge accumulation after electron movement in the conductor and have little effect on the field-induced electron emission process. The ambient gases in this study are mainly air with different concentrations. The gas composition, such as nitrogen or a vacuum environment, can significantly influence the discharge process of Townsend gas. The follow-up research plan will be the influence of different gas compositions on discharge parameters. The electrostatic discharge process concerned in this paper applies to most metal materials, including some precious metals with surface plasmon effect, such as gold, silver, and copper. Subsequent research plans will involve precious metals, graphene, and even metamaterials to explore the effect of surface plasmon resonance on discharge parameters excited by static electricity.

4. Analysis of environment factors with wavelet neural network

Based on Fourier analysis, wavelet analysis theory develops gradually. Wavelet analysis overcomes the limitation that the Fourier transform cannot reflect the relationship between frequency component and time variation when analyzing non-stationary signals and provides a time-frequency localization signal analysis method with fixed window size and variable shape. The scale scaling and time shift operations in wavelet analysis are very useful for analyzing instantaneous time-varying signals. The scaling operation can be used for signal refinement analysis. The time shift operation can be used to locate the signal period to be analyzed. Therefore, wavelet analysis can be used to extract information effectively from the signal and solve the problem that the Fourier transform cannot solve.

The time-frequency analysis of electrostatic discharge signal based on wavelet transform is applied to the case that the frequency changes with time in the process of the non-stationary signal. The information hidden in the electrostatic discharge signal can be obtained by time-frequency analysis. At the same time, the wavelet threshold denoising method is used to preprocess the ESD current signal, which can solve the noise interference problem in the current waveform collected in the test. The main idea of the wavelet denoising method is to use the wavelet analysis to decompose the signal in multiple layers, effectively distinguish the noise information in the signal, select the appropriate threshold and threshold quantization method, de-noising the high-frequency coefficient part, and obtain relatively stable observation data after the wavelet reconstruction. Then, the variance of unknown constant noise is estimated using Sage-Husa's Maximum a Posteriori (MAP) noise statistics estimator. The estimated value of the unknown noise variance is used as the modified information of the standard Kalman filter to modify the data after wavelet pretreatment further.

However, there are also some problems with the use of wavelets. Many kinds of wavelet functions are in different series, each with its characteristics, and many attempts are needed to select wavelet functions. For example, the wavelet functions of

Daubechies, Coieflet, Symlets, and Biorthogonal wavelets are selected as the mother wavelets to carry out multi-layer decomposition and threshold quantization of ESD current signal of superimposed noise. The experimental results show that the number of wavelet decomposition layers obviously affects the denoising performance. Generally, there is an appropriate number of decomposition layers (four to six layers), and the denoising effect gradually deteriorates when the number exceeds it. The order of wavelet function has a relatively weak influence on the denoising performance, and with the increase of wavelet order, the denoising performance will improve.

Wavelet bases can approach signals more accurately from a mathematical point of view and are more advantageous than Fourier transforms in signal processing. Wavelet transform has time-frequency local and focusing characteristics, while the neural network has self-learning, self-adaptation, robustness, fault tolerance, and generalization ability. Combining the advantages of the two has always been a concern of people. One method is to preprocess the signal with wavelet analysis. The wavelet space is used as the feature space of pattern recognition, and then the extracted feature vector is sent to the neural network for processing. The other is called wavelet neural network (WNN) or wavelet network (WN). Wavelet neural network was first proposed by Zhang Qinghu et al. from IRLSA, a famous French information science research institute, in 1992 [8]. Wavelet neural network is based on wavelet transform. It constitutes the neural network model, which USES nonlinear wavelet base replace neurons usually nonlinear excitation function (e.g., Sigmoid function), set the strongpoint of artificial neural network and wavelet analysis, even if the network convergence speed and avoid falling into local optimum, and sometimes the frequency characteristics of local analysis.

WNN is an artificial neural network based on the breakthrough of wavelet analysis. It is a new layered, multi-resolution artificial neural network model based on wavelet analysis theory and wavelet transform. WNN replaces the S function of the hidden node of the neural network with the wavelet function and the corresponding weight from the input layer to the hidden layer, and the threshold of the hidden layer is replaced by the scale scaling factor and time translation factor of the wavelet function, respectively. In other words, the nonlinear wavelet basis replaces the usual nonlinear sigmoid function, and the signal representation is expressed by linear superposition of the selected wavelet basis. It avoids the blindness of the BP neural network structure design and local optimal nonlinear optimization problems, greatly simplifies the training, has a strong function-learning ability and extension ability, and has broad application prospects. WNN has the following characteristics: First, the determination of wavelet elements and the entire network structure has a reliable theoretical basis, which can avoid blindness in the structure design of the BP neural network. Secondly, the linear distribution of network weights and the convexity of learning objective function make the network training fundamentally avoid the local optimal nonlinear optimization problem. Thirdly, they have strong function-learning ability and promotion abilities.

The process is conducting a linear simulation of an experiment with a linear neural network as the following: firstly, supplying input data and output data, then substituting the data and conducting circling training according to the establishment model, and lastly, determining prioritized value and the threshold value. Each neuron in the neural network is interconnected, and each neuron has two states, activated or suppressed, as input variables, corresponding to 1 or 0, respectively [6, 7]. The states of each neuron are governed by the following rules and interact within the overall neural network:

$$x_i = f\left(\sum_{j=1}^{N} w_{ij} x_j - \theta_i\right) \tag{3}$$

where w_{ij} is the weight of the association between two neurons i and j, that is, the strength of the synaptic connection.

θ_i is the neurons threshold.

$f(\cdot)$ -- $step(\cdot)$ is the of a step function, which contains

$$step(a) = \begin{cases} 1, a \geq 0 \\ 0, a < 0 \end{cases} \tag{4}$$

It is an effective tool to extract common neural networks using the MP model to analyze the discharge parameters of electrostatic discharge events, as shown in **Figure 3**. Specific mathematical expressions are shown in (5) to (7):

$$\sigma_i = \sum_{j=1}^{N} w_{ij} x_j + S_i - \theta_i \tag{5}$$

$$u_i = f(\sigma_i) \tag{6}$$

$$y_i = g(u_i) = h(\sigma_i) \tag{7}$$

The meaning of the formula w_{ij}, x_j, θ_i is the same as that in (3).

σ_i is the neuron input.

S_i is the external neuron input.

u_i is the neuron activation state.

y_i is the neuron output.

$f(\cdot)$ is the activation functions of neurons.

$g(\cdot)$ is the conversion functions of neurons.

The most important is the training process in the whole linear simulating process [6]. The solution training flowing sketch block diagram is given in **Figure 4**.

The neural network error in the training process is a multi-dimensional paraboloid. The best solution can always be a fund for a linear neuro-network in training based on the least square principle of gradient fall if the learning rate is low enough. A model can be established through previous analysis related to discharge parameters

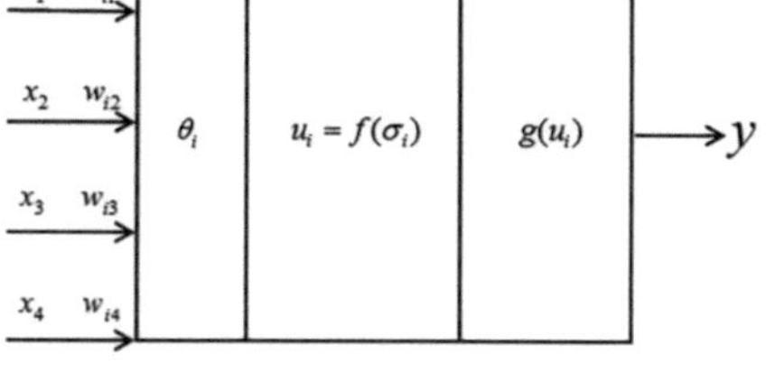

Figure 3.
MP model of neural network.

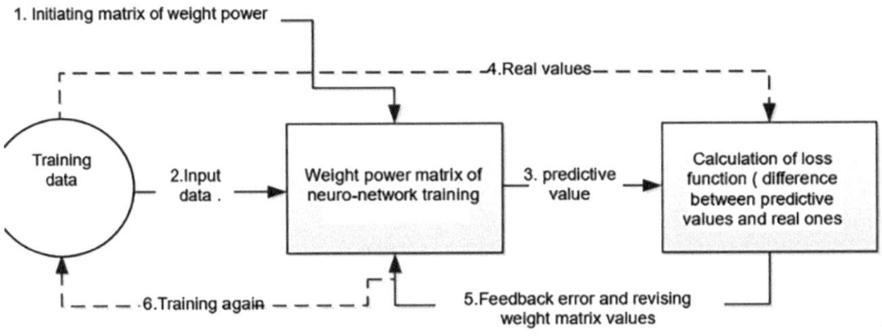

Figure 4.
Block diagram of solution process by linear training of neural network.

variation; the input variable and output variable can also be determined. The influence on discharge parameters by moving electrode speed is shown in **Table 2**.

It can be seen from **Table 2** that three output variables correspond to one input variable, which denotes that three linear fitting calculations will be performed separately. According to **Table 2** data, **Figure 5** shows the fitting diagram of the relationship between electrode approach velocity and discharge parameters. Peak current value increases with moving electrode speed lifting. The relation between the peak current of ESD and electrode moving speed can be seen distinctly through the linear fitting of the neural network. With the iteration number increase, as seen in **Figure 6**, the mean square error falls first, then becomes flat late, and reaches balance finally, which means the best-fitting straight line. A good agreement has been reached on analysis with two methods, that is, theoretical analysis and neural network analysis. That is to say, the consequence has been verified that a strong positive, relevant relationship exists between electrode moving speed and discharge peak values.

A wavelet transformer is a powerful tool for analyzing signals. The thought-suppressing noise is decomposing current signals with available wavelet function

Input	**Output 1**	**Output 2**	**Output 3**
Moving speed of electrode (m/s)	**Peak current** (A)	**Rise slope** (A/ns)	**Fall slope** (A/ns)
0.05	1.560	0.962	1.842
0.09	1.704	1.241	2.232
0.13	1.992	1.635	2.480
0.18	2.136	1.884	2.612
0.22	2.160	1.923	2.837
0.27	2.232	2.008	2.848
0.31	2.256	2.127	2.872
0.35	2.472	2.321	2.890
0.39	2.582	2.482	2.962
0.42	2.612	2.602	3.103
0.45	2.634	2.711	3.221

Table 2.
Linear model input and output data of ESD affected by electrode moving speed.

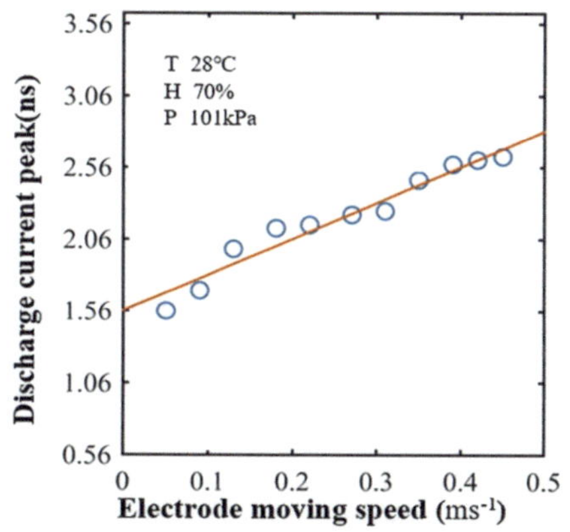

Figure 5.
Relationship of peak discharge current with electrode moving speed.

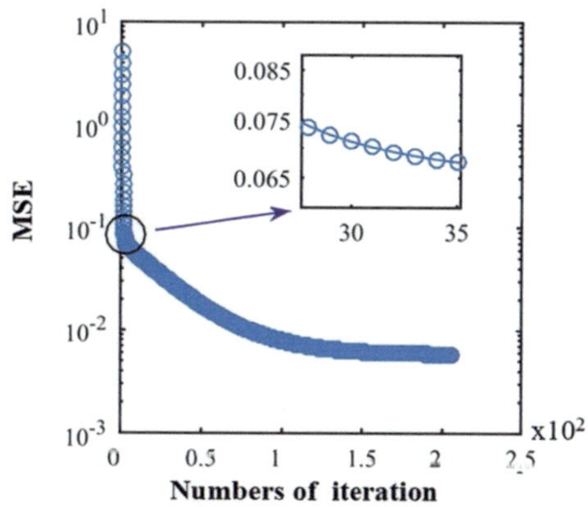

Figure 6.
Relation between iteration number and mean square error.

selection and decomposition layers [9, 10]. Information of outline on ESD current is mainly distributed in the low-frequency range, but the noise is distributed in the high-frequency range. So, noise in the ESD current waveform can be suppressed by removing high components in wavelet decomposition coefficients. The flowing block diagram is shown in **Figure 7**.

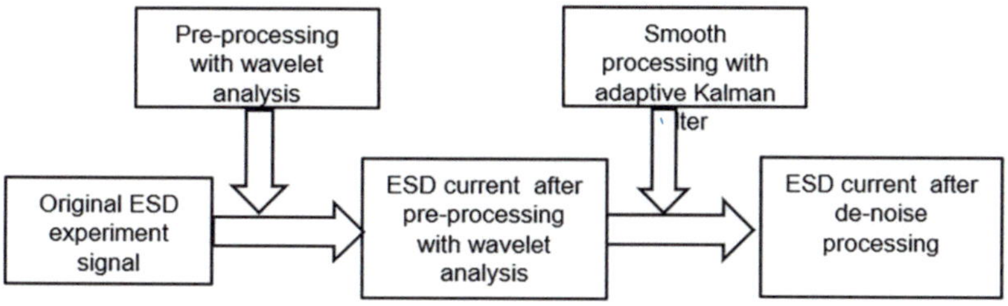

Figure 7.
Flowing block diagram with wavelet transform and Kalman filter method.

A good effect on noise suppression can be obtained in analyzing the current electrostatic discharge current waveform if the wavelet transformer is combined with the Kalman filter. After pre-suppressing with wavelet transformers and taking the adaptive Kalman filter algorithm, the electrostatic discharge (ESD) current waveform is shown to purify more fluent discharge waveform results.

5. Discrete characteristic analysis of discharge parameters

5.1 Correlation of discharge parameters with velocity to target

The multi-factor micro-gap electrostatic discharge test dominated by approach velocity was conducted at different approach speeds. The micro-gap range was 1 μm-1 mm, and the discharge current parameters showed regular changes, as shown in **Figure 8**. According to the analysis, under the condition of the same other influencing factors, such as the same discharge voltage, the higher the electrode approaching speed, the higher the discharge current peak value.

According to the human body/metal electrostatic discharge experiment [11], the electrode approach speed strongly correlates with discharge parameters such as current peak, peak current derivative, and arc length. The three discharge parameters have different correlation coefficients with electrodes approaching velocity at different voltages. The correlation between discharge parameters and electrode approach velocity is significantly different within a certain threshold of discharge voltage. **Table 3** shows the experimental test results of the correlation between discharge parameters and electrode approach velocity. According to the experimental data, when the charging voltage is equal to or less than 300 V, the electrode moving speed has almost no effect on discharge parameters. However, with the increase of charging voltage, the influence of electrode velocity is enhanced and reaches the maximum at about 800 V. When the charging voltage increases to more than 1500 V, the influence of moving electrode velocity on discharge parameters decreases. The mechanism of

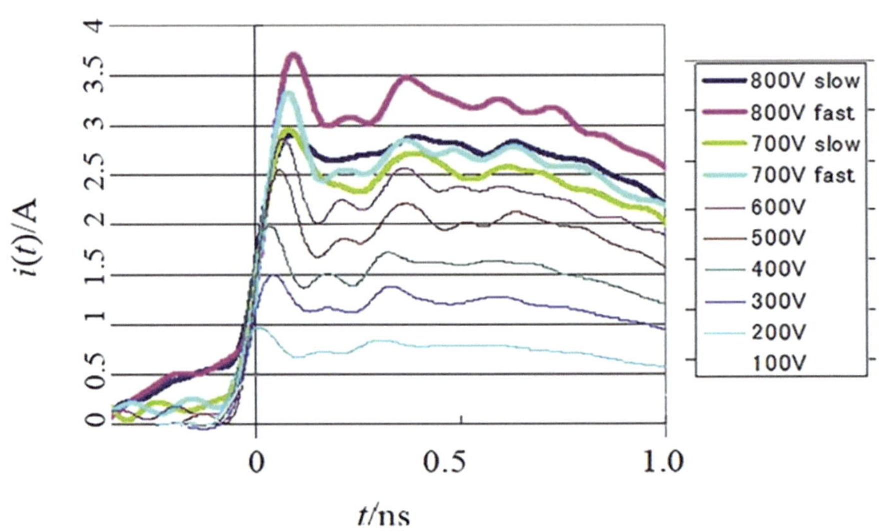

Figure 8.
Measured discharge current waveforms showing the strong influence of fast and slow electrode velocity.

$V_c[kV]$	Correlation coefficients					
	Peak current		Peak current derivative		Spark length	
0.3	0.029	NS	−0.022	NS	−0.028	NS
0.5	0.544	**	0.452	**	−0.532	**
0.8	0.668	**	0.644	**	−0.656	**
1.0	0.572	**	0.511	**	−0.516	**
1.1	0.555	**	0.557	**	−0.338	**
1.5	0.612	**	0.552	**	−0.482	**

*Note: **: much strong, *: strong, NS: no correlation.*

Table 3.
The influence coefficient of different voltages on discharge parameters.

the influence of approach velocity on discharge parameters is a research focus of electrostatic discharge, and the experimental results also verify the theoretical analysis of the two sub-processes of field-induced electron emission and Townsend gas discharge mentioned above.

5.2 Discharge parameters discrete characteristics

In an electrostatic discharge generator (ESD gun) experiment, the discharge current peak has significant discrete characteristics, even if the charge voltage is the same many times. **Figures 9** and **10** show the experiment discharge currents with an ESD gun. Two tests have a charge voltage of 6 kV. The ESD gun has a fixed discharge gap without electrode velocity in the first situation. **Figure 9** shows the waveform diagram of the discharge current. In another situation, the ESD gun was held in hand and moved slowly to the discharge target. **Figure 10** shows the corresponding discharge current waveform.

Figures 9 and **10** show that the discharge current peak values vary greatly. If an ESD gun has an identical charge voltage and the same gap, the discharge current usually should have the same discharge peak values, even if they have different waveforms. The results shown in **Figures 9** and **10** mean some other factors exist and strongly affect discharge parameters. Air pressure, relative humidity, and temperature

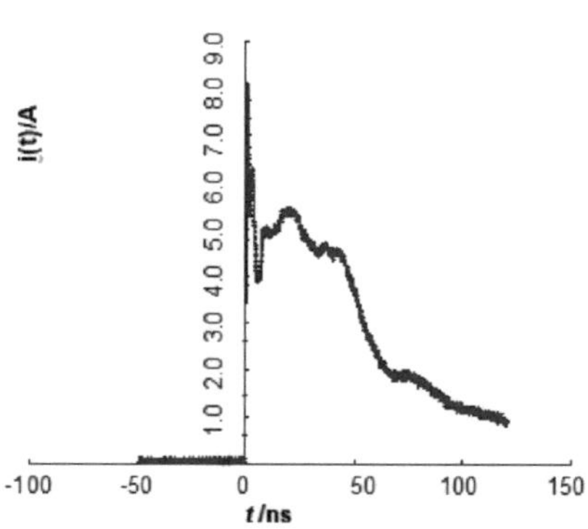

Figure 9.
Schematic diagram of discharging current at a faster approaching velocity.

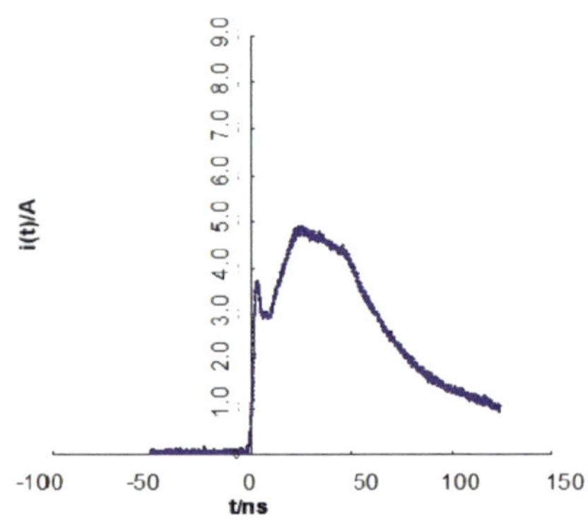

Figure 10.
Schematic diagram of discharging current at a slower approaching velocity.

may be the important factors resulting in discharge variation. Why and how discharge charges current peak values are correlated with these factors?

In two situations of the experiment, air pressure, relative humidity, and temperature have no obvious change as a matter of three factors. Discharge takes place in the first situation if the ESD gun is triggered. But, in the second situation, discharge occurs after the ESD gun is triggered and moved to the target for a certain time. Presumably, some of the charges that had accumulated in the metal tip of the discharge gun disappeared as it moved toward the target.

The reason why part of the charge disappeared in the process of electrode movement has been mentioned in Section II. After the discharge gun is charged, a high voltage difference is formed, and the movement of the electrode is more likely to produce gas discharge. As the gas was broken down and ionized, it suddenly changed from an insulator to a good conductor, and the charge was transferred, and the current surged, so part of the charge disappeared. However, the power of the power supply is not enough, so the voltage drops, the discharge is temporarily stopped, and the discharge is again after the voltage is restored, so the spark discharge has a discontinuity.

Differences in discharge current peak values in our experiment vary with charge voltage in two situations with an ESD gun. The difference in peak current at a charge voltage of 2 kV is quite small. At 4 kV, however, the difference increases distinctively and gets drastically larger at 6 kV. The difference at 6 kV could reach more than one-half of the total value of the discharge current.

6. Conclusion

The influence of electrode proximity velocity on discharge parameters exists in micro-gap electrostatic discharge, which can explain the mutual transformation of two electron-discharge processes, field-induced electron emission and Townsend gas discharge at a low voltage of 0.8 kV–1.0 kV. Experimental verification based on a wavelet neural network agrees with the above theoretical analysis, which explains the influence of electrode proximity velocity on discharge performance in micro-gap electrostatic discharge and reveals the discrete characteristics of discharge parameters. We will discuss the relationship between the loss of charge and three factors: air pressure, relative humidity, and temperature. The research on the mechanism of micro-gap electrostatic discharge dominated by proximity velocity is of great practical significance, not only in the electrostatic protection of micro-electronic devices and systems but also in the formulation of immunity test standards.

Acknowledgements

I would like to express my sincere thanks to Prof. Ruan Fangming from Guizhou Normal University, Dr. Chen Ze from Yancheng Normal University, and Prof. Meng Cui from Zhejiang University for their significant assistance in the completion of the manuscript.

Conflict of interest

The authors declare no conflict of interest.

Author details

Yang Meng[1*], Fangming Ruan[2], Yan Lyu[3], Ze Chen[4], Dandan Qu[1] and Kun Zhang[1]

1 Chinese Academy of Sciences, Institute of Mechanics, Beijing, China

2 School of Big Data and Computer Science, Guizhou Normal University, Guiyang, China

3 School of Health and Aged Care, Shandong Women's University, Jinan, China

4 School of Physics and Electronic Engineering, Yancheng Teachers University, Yancheng, China

*Address all correspondence to: mengyang@imech.ac.cn

References

[1] Daout B, Ryser H, Germond A, Zweiacker P. The correlation of rising slope and speed of approach in ESD tests. In: Proc. 7th Int. Symp. Electromagnetic Compatibility. Zurich, Switzerland: IEEE; 1987. pp. 461-466

[2] Pommerenke D. "On the influence of the speed of approach, humidity and arc length on ESD breakdown", ESD Forum, Grainau, Germany: É" kiuanea of the Speed of Approach, Humidity and Arc Length on ESD breakdown; 1993. pp. 103-111

[3] Pommerenke D. ESD: Transient fields, arc simulation, and rise time limit. Journal of Electrostatics. 1995;**36**:31-54

[4] Pommerenke D. Martin, Aidam, "ESD: Waveform calculation, field and current of human and simulator ESD". Journal of Electrostatics. 1996;**38**:33-51

[5] Sumida A, Yoshida T, Masui N. Effects of the relative humidity on ESD from the charged metal. Seidenki Gakkai Koen Ronbunshu. 2006;**2006**:43-44

[6] Bonish S, Kalkner W, Pommerenke D. Modeling of short-gap ESD under consideration of different discharge mechanisms. IEEE Transactions on Plasma Science. 2003;**31**:736-744

[7] Xue ZQ, Wu QD. Electronic Emission and Electronic Spectrum. Beijing: Press of Beijing University; 1993. pp. 62-75

[8] Zhang Q, Benveniste A. Wavelet networks. IEEE Transactions on Neural Networks. 1992;**3**(6):889-898

[9] Mori I, Fujiwara O. Severity Evaluation of Immunity Test for Air and Contact Discharges of ESD Generators, Proceedings of CEEM2009, Xi'an; 2009. pp. 110-113

[10] Ruan F, Gao Y, Shi D. Analysis on Electrode Speed Correlation of Discharge Parameters Applying Short-gap Electrostatic Discharge Model. In: Proc. IEEE International Conference on EMC. Detroit, Michigan, U.S.; 2008. pp. 1-4

[11] Ruan F, Fujiwara O. Effect of approaching contact speed of hand-held metal piece on characteristics of discharge current from charged human body. In: Proc. the 4th Asia-Pacific International Conference on EMC (CEEM'2006). Dalian, China: IEEE; 2006. pp. 107-112

Chapter 3

Approximation of ESD Protection on Automotive Headlight Production Lines

Michal Sahul and Peter Janiga

Abstract

The use of LED technology in headlights has made the manufacturer susceptible to potential damage due to electrostatic discharge (ESD). This chapter focuses on the risks of ESD and ensuring manufacturing compliance with IEC 61340-5-1 or ANSI/ESD S20.20 standards. The first section will provide a theoretical introduction to the protection of ESD-sensitive components. It describes the sources and mechanisms of ESD in the automated production of automotive headlights, the risk situations in production lines, and the protective measures to eliminate this phenomenon. Then, some of the measurement methods used to identify ESD protection parameters performed on-site are described. In the next part of the chapter, a case study based on a typical production line for automotive headlights is described. The case study describes the different workstations within the line in terms of processes and the probability of occurrence of electrostatic potential, together with the design of ESD protective measures for the entire production line, taking into account known sources of ESD risks.

Keywords: ESD control program, ESD sensitive device, ESD risk analysis, ESD risk, human body model, charged device model, isolated conductor, ESD protective measure, ESD measurement

1. Introduction

With the development of LED technology in automotive headlights and the concomitant increase in requirements for maintenance-free, reliable, and durable headlights, headlight production has become highly sensitive to potential damage due to electrostatic discharge (ESD). Since LED technology and its control in the headlight contain a PCB (printed circuit board) with many electrostatic discharge sensitive device (ESDS) components, it is necessary to implement ESD protection and to arrange the production in accordance with the latest requirements of standards such as IEC 61340-5-1 or ANSI/ESD S20.20 [1]. ESD protection requirements introduced in manufacturing companies through an ESD control program adapt the processes of manufacturing, assembly, installation, packaging, marking, testing and inspection, transportation, and handling of ESDS [1]. An ESD control program based on the above-mentioned standards mainly focuses on the administrative and technical requirements of ESD protection [2]. The purpose of this protection is to set aside an ESD protected area (EPA) and, through measures to prevent damage to ESDS, reduce the risk to a standard acceptable minimum or eliminate it altogether. The risk of ESD damage to components must be realized before a production line or working machine is put into operation. Risk management is an integral part of every supply chain aspect [3]. Unsuccessful and incomplete risk management can have negative economic consequences.

This chapter is based on the requirement for the design of ESD protective measures of a production line for assembling automotive headlights. It describes the theoretical background necessary to understand ESD protection and how to implement it. This section describes the requirements for an ESD control program, on which the procedure for designing ESD protection measures is based. The chapter also includes a selection of basic measurements performed as part of the verification of compliance with the control program and a description of ESD protection models. In the last section, a case study of a simple production line model is presented with the design of ESD measures based on the assumed ESD threats.

2. Theoretical foundations of ESD protection for sensitive components

As previously mentioned, ESD protection has become an essential part of most manufacturing processes where electrostatic sensitive devices (ESDS) are handled. In terms of production processes, any component, product, or part of a product can be considered ESDS if it is susceptible to damage from the phenomenon of discharging accumulated electrostatic energy from a person, insulator, or conductor. According to one of the fundamental technical standards for ESD protection [4], electrostatic discharge is a serious threat, especially to electronic components. The most sensitive ESDS, in terms of potential damage, include semiconductors, thin-film resistors, and similar components. Due to the high sensitivity of electronic components, it can be assumed that any discharges occurring during the handling of ESDS in production could be a source of partial or complete damage to the component.

2.1 ESD protection in general

In ESD protection, it is generally accepted that the value of electrostatic potential capable of destroying a component is significantly lower than the human body's sensitivity threshold. The presence of electrostatic potential can, therefore, only be detected through proper measurement. For this reason, ESD protective measures in manufacturing must be set up preventively, focusing on the production process.

When protecting ESDS, it is important to consider situations where electrostatic charge may occur. In general, a charge is generated or accumulated on the surface of a material or product. Damage occurs when an electrostatic discharge happens from the surface of one material to another, with the ESDS being part of this circuit or exposed to induction.

2.1.1 Triboelectric charging (contact charging)

A significant source of charging is the so-called triboelectric charging. When two initially uncharged surfaces come into contact with each other, charge transfer occurs at their common interface [4]. After these materials are subsequently separated, a mutual electrical imbalance remains on their surfaces, which can be defined as an electrostatic charge, electrostatic potential, or an electrostatic field. The following **Figure 1** illustrates the charging of two insulating materials as a result of the triboelectric phenomenon. Step 1 shows the contact between two uncharged materials, and Step 2 illustrates the electrostatic state after they are separated.

The two materials are oppositely charged, resulting in an electrostatic field between them. As these materials move apart, work must be done to overcome the attractive forces of the opposite charges, and the potential difference increases linearly with distance. If these objects were made of conductive material, recombination would occur, leaving no significant amount of charge on their surfaces. However, if the objects are made of dielectric material ($R > 10^{11}\ \Omega$), full recombination does not occur, and a substantial amount of charge remains on the surfaces.

Depending on the type of material, the size of the contact area, and the distance between them, the intensity of the resulting electrostatic field can vary. In practice, charging can reach several tens of kV/m [4].

2.1.2 Charging by electric induction

In manufacturing processes, a problem arises when an ESDS (Electrostatic Sensitive Device) enters an existing electrostatic field. The risk principle here is based on electrical induction. When an electrostatic field exists around a charged object (for example, inside the plastic housing of an automotive headlight), any conductive or dissipative material within that field will be subjected to electrical induction. If the conductor is not grounded, it will acquire a potential due to electrical induction that depends on its position within the electrostatic field [4].

In practice, a scenario might occur where a PCB is installed into a plastic housing. This housing could be charged due to a triboelectric effect (**Figure 2**). During the installation of the PCB, electrical induction will manifest on its conductive parts, leading to the charging of these conductive parts of the ESDS based on the intensity of the electrostatic field (**Figure 3**). Even if the ESDS is not damaged

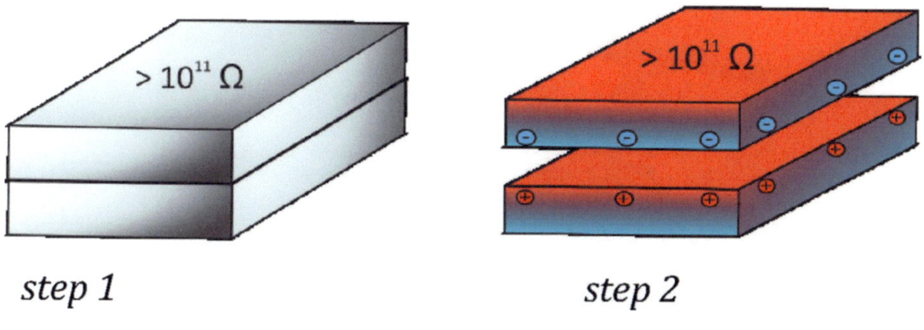

Figure 1.
Triboelectric charge of two dielectric surfaces.

under these conditions, there is still a risk of damage if it comes into contact with a grounded conductive part or person. The resulting electric current from the potential difference between the charged part and another conductive part can cause damage to the ESDS.

2.1.3 Classification of materials in ESD protection

In the principles of ESD protection for sensitive components, it is necessary to classify materials based on their conductivity differently than in conventional electronics. Simply put, for ESD protection, materials are classified according to their ability to dissipate electrostatic charge from their surfaces. This ability is related to the electrical conductivity of the material, which also affects its resistance to triboelectric charging.

The properties of materials are utilized in all manufacturing applications where ESDS are handled. For example, one can mention:

- packaging,

Figure 2.
Charged housing of automotive headlight—measured 17.6 kV/m.

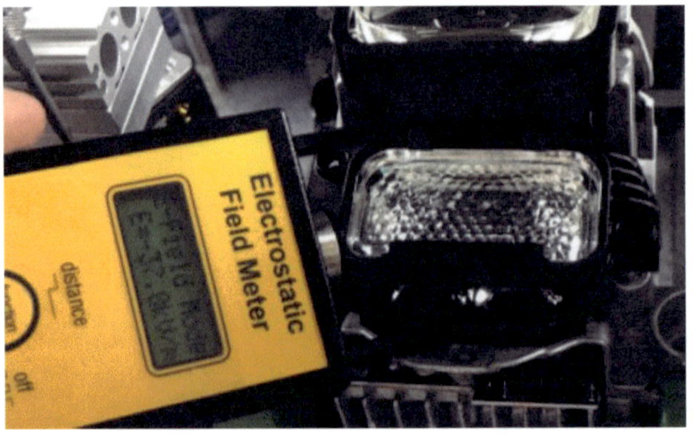

Figure 3.
Charged plastic parts of automotive headlight with PCB installed—measured 37.0 kV/m on a PCB.

- work surfaces,
- materials used near ESDS handling (e.g., polycarbonate in workstations),
- tools and other components used during ESDS handling, and so on.

Figure 4 illustrates the general classification of materials used in ESD protection. This classification of materials based on resistive properties is derived from the classification of ESD packaging materials [5].

2.2 ESD protection in manufacturing company

In a manufacturing company, ESDS (Electrostatic Sensitive Devices) must be systematically protected. This protection system is typically established within controlled documentation as an internal ESD control program. This program should be designed and tailored according to the manufacturing process, personnel, sensitivity of ESDS, production methods, and other factors. The protection of sensitive components in a manufacturing facility is managed by designated personnel who ensure regular inspections and compliance with established requirements. To this end, protected work areas known as EPA (ESD Protected Areas) are created, measures are implemented, regular measurements are conducted, and personnel training is provided. **Figure 5** illustrates the general ESD protection symbol used worldwide.

2.2.1 Internal ESD control program

This document must address all requirements related to the handling of ESDS (Electrostatic Sensitive Devices). Generally, it pertains to the manufacturing, processing, assembly, installation, packaging, labeling, servicing, testing, inspection, and transportation of ESDS both within and outside the company [1]. The document must

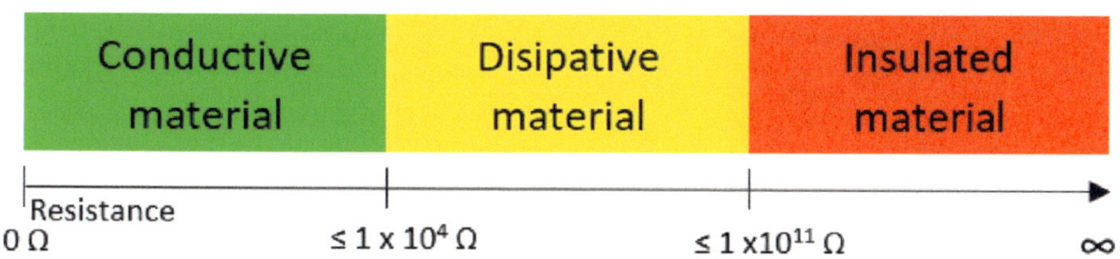

Figure 4.
General classification of materials in ESD protection according to resistance properties.

Figure 5.
Graphic symbol of ESD protection marking [6].

consider these processes and establish both administrative requirements and technical requirements. **Table 1** provides a summary of these requirements.

All items used for ESD protection of components fall under the internal control program. The implementation and regular verification of compliance are the responsibility of a designated person—typically an ESD Coordinator [1]. The organization must appoint this person or a team to coordinate ESD protection. The ESD Coordinator is responsible for ensuring that the requirements of the established control program are met and generally reports to the company's management. In the event of customer audits, the ESD Coordinator is accountable for any discrepancies found and their resolution.

2.2.2 ESD protected area—EPA

This designated and properly marked area is used for the direct handling of ESDS (Electrostatic Sensitive Devices). If an ESDS is being removed from its protective packaging, it must be done within the EPA (ESD Protected Area). As previously mentioned, this area is characterized by ensuring continuous grounding of all conductive or dissipative ESD protection items. All entrances must be marked with a warning that an individual is entering the EPA. The EPA can be established as a standalone workstation, a room, or a section of the production area. **Figure 6** illustrates the setup of an EPA workstation in an island configuration within a manufacturing facility.

Typical ESD protection items in an EPA (ESD Protected Area) include:

Administrative requirements	
Requirement	**Description**
Training plan	The content also addresses the training plan for staff, who must be aware of the risks associated with ESD. The goal is to have sufficiently informed personnel in the company who contribute to effective ESD protection through their knowledge.
Product qualification plan	The organization must appropriately qualify the suitability of all items used for ESD protection. Product qualification is usually carried out before the item is introduced into the system.
Compliance verification plan	The organization must periodically verify the effectiveness of ESD protection items. Compliance verification is usually performed through visual inspection and measurement of ESD parameters.
Technical requirements	
Bonding and grounding systems	They address the methods of grounding and equipotential bonding of ESD protection items in order to prevent potential differences between ESDS and other conductive/dissipative materials with which it may come into contact.
Personal grounding	It addresses the grounding method for personnel handling ESDS. Personnel grounding is typically achieved through ESD footwear and flooring, which must meet normative requirements, or through the use of an ESD wrist strap assembly.
EPA requirements	It addresses the arrangement of EPA workstations. Each workstation must ensure ESD protection of the production process, considering the technology, type, and susceptibility of ESDS.
Packaging	It addresses the packaging program designed for the storage and transport of ESDS. The packaging program must meet the requirements of the customer/supplier and must have the correct labeling according to the classification of the packaging material.
Marking	It addresses the labeling method for ESDS, workstations, and packaging materials to ensure compliance with customer/supplier requirements.

Table 1.
Brief overview of administrative and technical requirements of the ESD control program [1].

- Grounded work surfaces—Typically made using ESD mats with dissipative surfaces.
- Grounded storage shelves—Generally equipped with ESD mats with dissipative surfaces.
- Grounded transport carts—Usually grounded through the floor via ESD wheels.
- Grounding points for wrist straps—To ensure proper grounding of personnel.
- ESD flooring—Commonly made from poured epoxy floors, adhesive PVC floors (**Figure 6**), or portable ESD mats.
- Ionization—To neutralize static charges on non-conductive surfaces.
- Seating—Typically grounded through the floor via ESD wheels.
- Clothing—Provides protection for ESDS from the electrostatic field generated by the worker's clothing. ESD clothing can be directly grounded or non-grounded.

2.2.3 Isolated conductors and ionizing

If there are isolated (non-grounded) conductors near or in direct contact with ESDS (Electrostatic Sensitive Devices), it is necessary to ensure that the potential difference between the ESDS and the isolated conductor does not exceed 35 V. This situation might occur, for example, if a PCB is inserted into a charged housing along with a heatsink, and a voltage is induced on the metallic heatsink. This risk can be mitigated by removing the source of charging, but this is not possible if the charging source is part of the product. In such cases, ionization can be used.

An ionizer is a device that generates and introduces ions with a positive charge (missing electrons) and a negative charge (excess electrons) into the charged area. Ions are typically generated from oxygen or nitrogen molecules [6]. The most

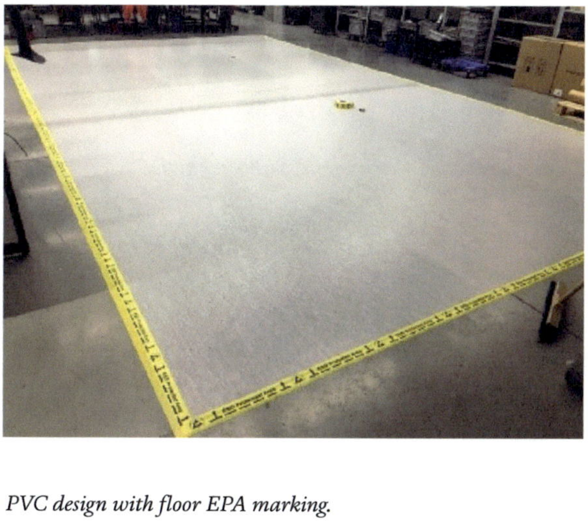

Figure 6.
Bonded ESD flooring in PVC design with floor EPA marking.

common ionization mechanisms are based on high-voltage corona discharge sources or, under certain conditions, radioactive sources [6].

- The Principle of High-Voltage Corona Discharge Source (**Figure 7**). This principle involves creating an electric field with an intensity exceeding 3 MV/m (at atmospheric pressure), which causes electrons to be released from the atomic shells.

- The Principle of Radioactive Source (**Figure 8**). This typically involves polonium-210. Alpha particles emitted by nuclear ionizers (positively charged helium nuclei—two protons, two neutrons, and no electrons) collide with air molecules, causing electrons to be ejected from some air molecules. This results in the formation of positive ions. When free electrons are captured by other air molecules, negative ions are formed [6].

By introducing ions into the charged space, recombination occurs, which neutralizes the resulting electrostatic charge. The balance of the ionizer (±) should be adjusted evenly to prevent the space from becoming recharged. The normative requirement for residual voltage is the same as the maximum voltage on isolated conductors - 35 V. **Figure 9** illustrates the ionization of an assembly fixture using a blow-off ionizer.

2.2.4 Simulation models of electrostatic discharges (ESD models)

From the perspective of coordinating ESD protection within a company, it is advantageous to understand the sensitivity of ESDS (Electrostatic Sensitive Devices) to electrostatic discharges. The manufacturer of ESDS can subject the component to

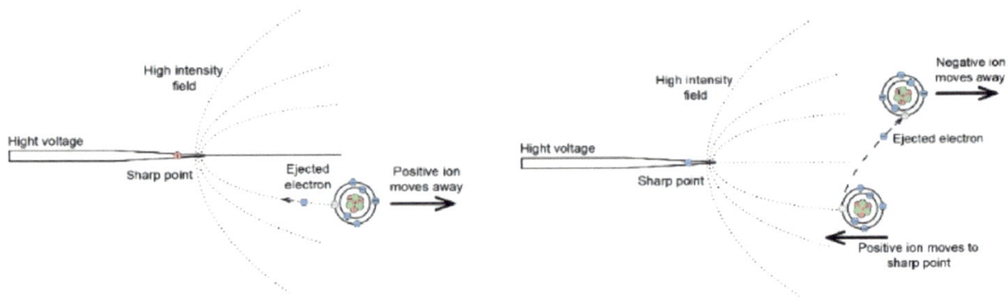

Figure 7.
Principle of high-voltage corona ionization (left-positive, right-negative).

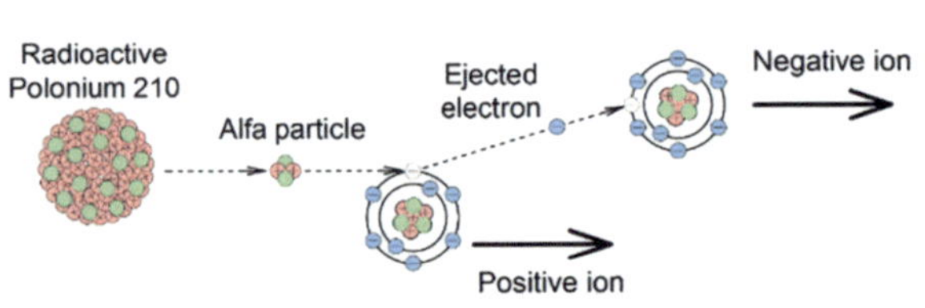

Figure 8.
Principle of a radioactive source ionization.

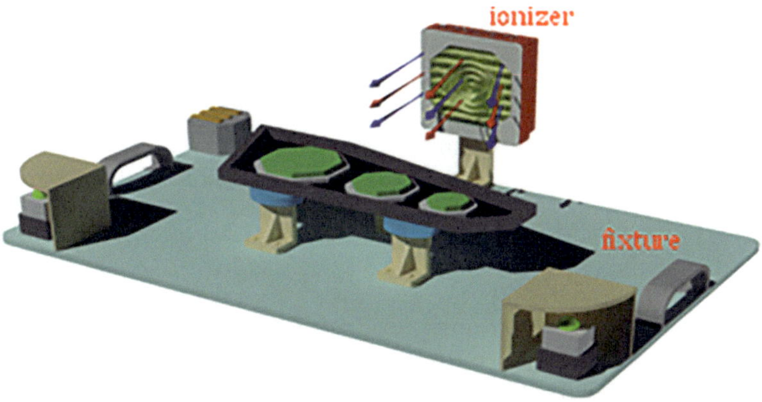

Figure 9.
Example of ionization of an assembly fixture when installing ESDS into the insulated housing of an automotive headlight.

electrostatic discharge simulation tests during development. Electrostatic discharge can be simulated using a simple circuit where an electrostatic charge is initially stored in a capacitor. The simulation involves discharging the capacitor into a load (the test component) while monitoring the current flow. This model can generate different current profiles depending on the values of the capacitor, resistor, and inductance [4]. **Figure 10** illustrates a basic circuit diagram for simulating electrostatic discharges to ESDS.

The values of capacitance, resistance, and inductance are generally adjusted according to three basic simulation models that correspond to risk scenarios that may occur during the manufacturing process.

- *Human body model (HBM)*: The model where ESDS is subjected to electrostatic discharge from the human body to a grounded ESDS [7]. The risk associated with manufacturing automotive headlights is significant, as manual workstations on production lines involve operators handling ESDS directly.

  Standard sources for testing ESDS in the HBM model specify basic parameters: R = 1000–3000 Ω; C = 100–300 pF; L is limited to a parasitic value [4].

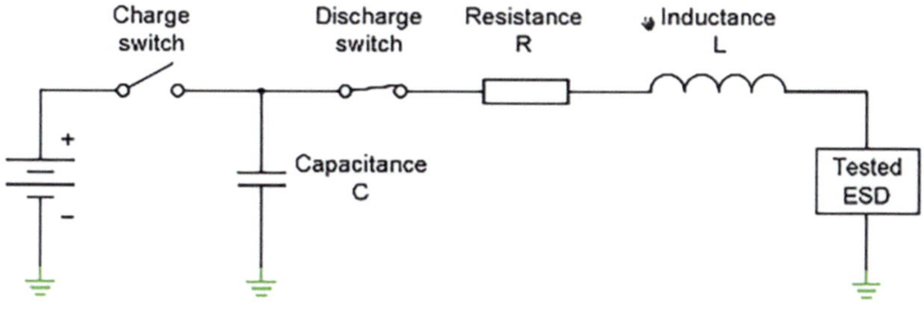

Figure 10.
Basic circuit diagram for simulating electrostatic discharges into ESDS.

- *Charged device model (CDM)*: The model where ESDS is charged during the process (e.g., due to triboelectric charging) and faces a so-called hard discharge into a grounded conductive material. The risk in manufacturing automotive headlights is also significant here. ESDS are assembled and connected into insulated plastic housings, where triboelectric effects and high electrostatic fields can occur.

  Standard sources for testing ESDS in the CDM model specify basic parameters: $R = < 10\ \Omega$; $C = 3\text{–}30$ pF; $L < 10$ nH [4].

- *Isolated conductors (IC)*: Previously used as the Machine Model (MM). In the case of ESDS contacting an electrically conductive part that is not grounded (e.g., metal), a discharge can occur from an isolated charged conductor to a grounded ESDS. In the manufacturing of automotive headlights, ICs are primarily found as metal components such as heatsinks, cable harness connectors, and similar parts [8].

  Standard sources for testing ESDS in the MM (IC) model specify typical parameters: $R = < 8.5\ \Omega$; $C = 200$ pF; $L < 0.5$ nH [4].

The use of electrostatic discharge simulation models is also valuable for assessing the risk of ESDS damage, for example, during the implementation and operation of a production line. By using these models, it is possible to evaluate the ESD risk at individual workstations on the production line [8].

3. Overview of measurement methods in EPA when verifying compliance in the company

As part of the coordination of ESD protection, the so-called compliance verification with the ESD control program. As part of the conformity verification process, it mainly concerns the performance of measurements according to the prescribed methods [9] and with measuring instruments subject to the calibration regime. The measured results must be in accordance with the normative document and the internal control program [4]. The following articles provide examples of the most frequent measurements performed in compliance verification. According to IEC 61340-5-1, measurements focused on personnel grounding and measurements focused on ESD protection items must be performed in the EPA area.

3.1 Methods of measuring personnel grounding

As previously mentioned, only individuals who meet grounding requirements should be present in the EPA. A fundamental requirement is ESD footwear, which provides continuous discharge of charge from the worker's body through the ESD floor to the ground. Compliance is typically verified in two ways:

- Measurement of Resistance to Groundable Point (Rgp) in person-footwear system [10]

- Measurement of Resistance to Ground (Rg) in footwear flooring system [1]

3.1.1 Measurement of resistance to groundable point (Rgp) in footwear flooring system—Footwear personal ground test

This measurement is typically performed before a person enters the EPA and is conducted by the individuals themselves. The measurement is carried out using a device called a "personal grounding tester (PGT)," which is designed for self-use by personnel. The person stands on the measuring electrode while wearing ESD footwear and presses a contact area on the PGT checker with their hand. The measurement is done with 9 and 100 V DC and the resistance to the measuring electrode (the contact area of the footwear with the electrode) must meet the threshold value of Rgp $< 1x10^{8}$ Ω. The PGT may also be equipped with accessories for measuring wrist straps. **Figure 11** illustrates an example of a PGT installation before entering the EPA. In this case, the PGT is synchronized with a turnstile that grants access to the EPA only based on a positive test result.

3.1.2 Measurement of resistance to ground (Rg) in footwear flooring system

This measurement is performed at regular intervals (according to the control program) on the installed flooring in the EPA. The goal of the measurement is to evaluate the ability to continuously discharge electrostatic charge through a sufficiently low resistance between the person and the ground. The principle involves measuring resistance with a voltage ranging from 10 V to 100 V DC. The measurement electrode is held by the person, and the resistance measurement is conducted relative to the ground [10]. Resistance in the footwear flooring system must meet the requirement of *R*g $< 1x10^{9}$ Ω and a maximum of 100 V of body voltage generation (**Figure 12**) [1, 10].

3.1.3 Measurement of body voltage generation on flooring system

This measurement is performed at regular intervals (according to the control program) on the installed flooring in the EPA. The goal of the measurement is to evaluate the ability to continuously discharge electrostatic charge through sufficiently low

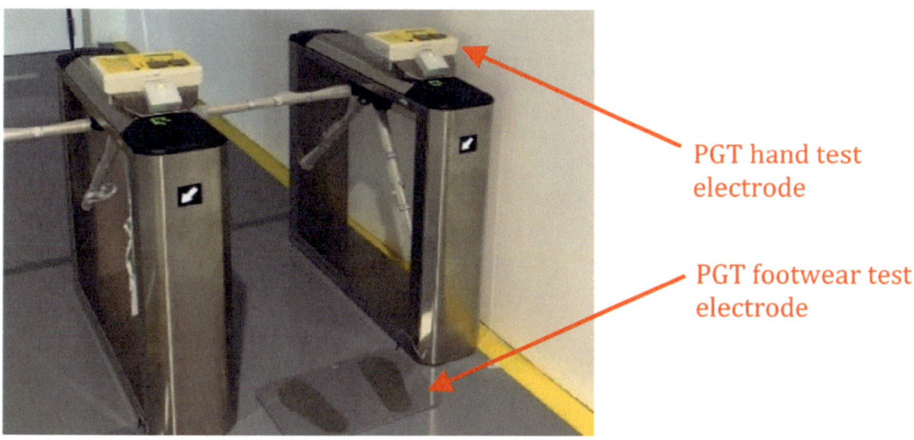

Figure 11.
Example of placement of PGT before entering the EPA in combination with a turnstile.

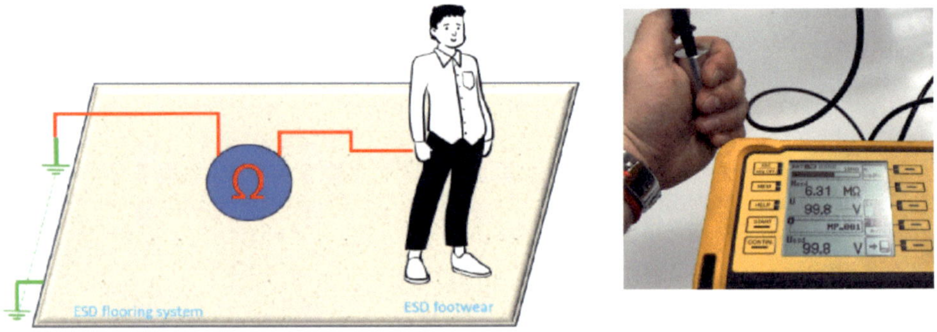

Figure 12.
Measurement of resistance to ground in footwear flooring system.

resistance between the person and the ground. The principle involves measuring the voltage generated by the person during walking (walking test) and the flooring's ability to discharge accumulated energy. The measurement electrode is held by the person, and the voltage during walking is measured relative to the ground [10]. Body voltage measurement must meet the requirement of a maximum voltage of 100 V, which is the calculated average of the 5 highest measured values (**Figures 13** and **14**) [1, 10].

3.2 Methods of measuring ESD protection items

The following sections outline the most commonly used measurement methods for ESD coordination to verify compliance. ESD protection items (ESD items) are measures used in the EPA (see 2.2.2) and must also undergo regular inspections and measurements [1]. The measurements detailed in the following sections represent a selection of the most frequently used methods for verifying compliance. However, many other measurements are employed in product qualification or laboratory evaluations.

3.2.1 Point-to-point resistance measurement

Point-to-point resistance measurement (Rpp) is a test used to determine the resistance value on the surface of a material. The measurement is performed using

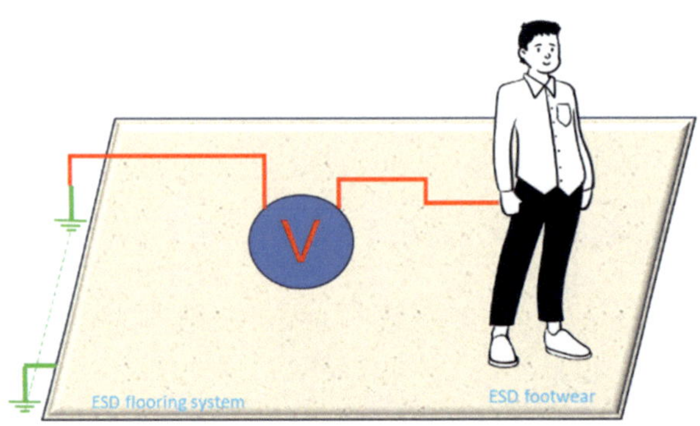

Figure 13.
Measurement of body voltage generation on flooring system.

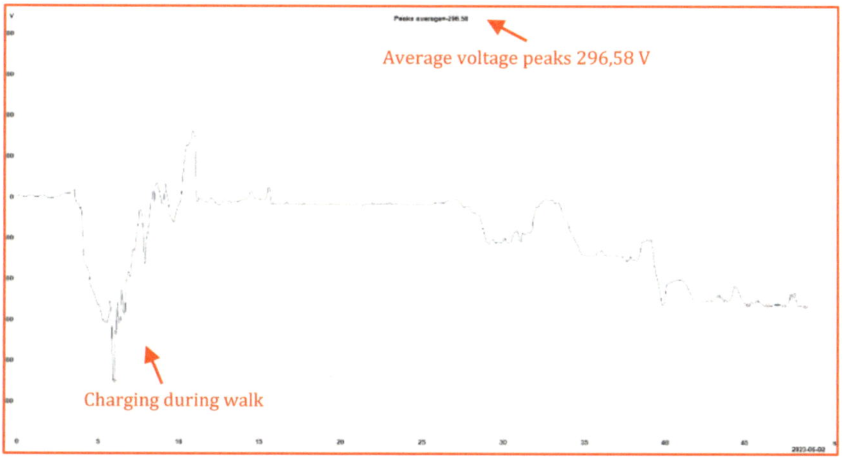

Figure 14.
Graphical progress and result of walking test measurement.

a resistance meter with a voltage of 10 V DC (± 0.5 V) for ranges up to 1 x 10^6 Ω, or 100 V DC (± 5 V) for ranges equal to or greater than 1 x 10^6 Ω [11].

This method is commonly used in practice to measure working surfaces, floors, clothing, and similar items. The resistance is measured between two electrodes placed on the surface of the material at a distance of approximately 25 cm [11]. The electrode used in this measurement typically weighs 2.5 kg (± 0.25 kg) and has a contact surface diameter of 63.5 mm (± 1 mm) (**Figure 15**) [11].

3.2.2 Resistance to ground measurement

Resistance to ground (*R*g) measurement is a test used to determine the resistance of the "lead path" between the surface of the material and the ground. This measurement is performed on installed ESD protection items and evaluates the entire grounding path. The measurement is conducted using a resistance meter with a voltage of 10 V DC (± 0.5 V) for ranges up to 1 x 10^6 Ω, or 100 V DC (± 5 V) for ranges equal to or greater than 1 x 10^6 Ω [11].

This method is commonly used in practice to measure the grounding path from work surfaces, floors, transport carts, grounding points for wrist straps, seats, grounded clothing, and similar items. The resistance is measured between a single electrode placed on the surface of the material and the ground (**Figure 16**) [11].

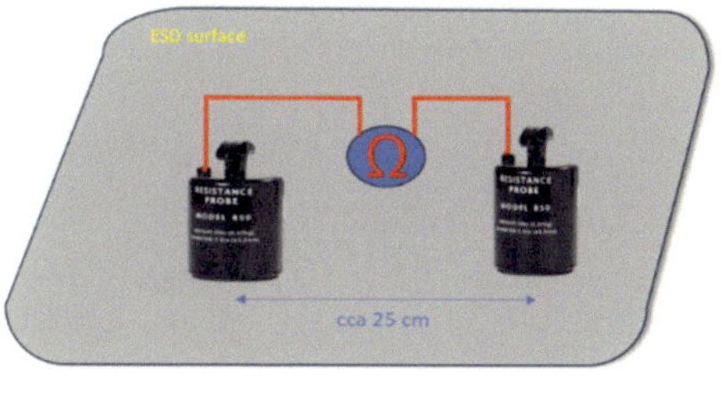

Figure 15.
Point-to-point resistance measurement on a surface (Rpp).

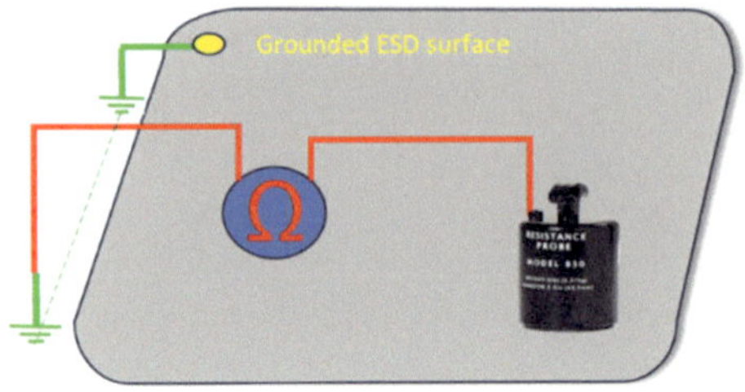

Figure 16.
Resistance to ground measurement (Rg).

3.2.3 Resistance to groundable point measurement

Resistance to groundable point (Rgp) measurement is essentially the same as resistance to ground measurement. The difference lies in the reference point against which the measurement is taken. This measurement is commonly performed, for example, when an ESD item is installed but not yet connected to an existing grounding system. This method can be used for partial measurements of unconnected floors, and work surfaces, as well as transport carts, seats, and similar items (**Figure 17**) [11].

3.3 Measurements on insulators in production

As previously mentioned, insulators that are prone to triboelectric charging and cannot be discharged through galvanic grounding are present in manufacturing environments. The ESD Coordinator must address these insulators in proximity to ESDS. ESD protection theory classifies insulators into two groups [1]:

- Nonessential insulators are those not directly related to production, and their presence near ESDS is unnecessary. These include personal items, various packaging, and items unrelated to ESDS. Such nonessential insulators can become highly charged and pose a risk to ESDS safety. These unnecessary insulators should be removed from the workspace, and the workspace should be organized to prevent charging risks. If they are part of the workspace (e.g., PC monitors, measurement devices, etc.), they should be arranged in a way that minimizes the risk of ESDS damage. Typically, this is achieved by placing them at a sufficient distance from ESDS. Sufficient distance is ensured when measurements of the electrostatic field or potential do not exceed normative parameters [1].

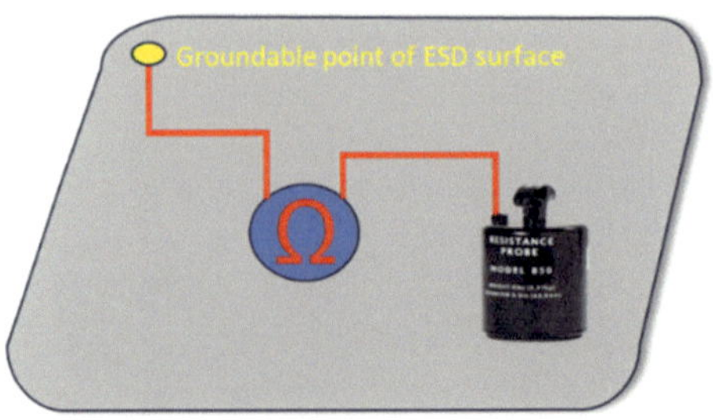

Figure 17.
Resistance to groundable point measurement (Rgp).

- Essential insulators** are those necessary for the processing and production of the product (e.g., automotive headlight housing, cable harnesses, other plastic components, etc.). If these insulators generate excessive electrostatic field values or potentials, there is a risk of damaging ESDS. In such cases, ionization must be used [1].

3.3.1 Electrostatic field intensity measurement

Measurement is conducted with an electrostatic field meter at the location where ESDS is handled. This measurement provides information about the intensity of the electrostatic field and helps determine the need for additional measures regarding the insulator. In this context, it is stipulated that the electrostatic field at the location where ESDS is manipulated must not exceed 5000 V/m. **Figure 18** shows a measurement of the electrostatic field intensity of 18.4 kV/m near a polycarbonate cover (left) and 0 kV/m after the removal of the polycarbonate (right).

3.3.2 Electrostatic potential measurement

Measurement is conducted with an electrostatic field meter (or another appropriate measuring instrument in potential measurement mode) on the surface of the insulator at a certain distance from the ESDS handling area. This measurement provides information about the surface potential of the insulator and determines the minimum distance from the ESDS. If the distance cannot be maintained, ionization must be used. In this context:

- If the electrostatic potential on the surface of the insulator exceeds 2000 V, the ESDS must be at least 30 cm away from the insulator.

- If the electrostatic potential on the surface of the insulator exceeds 125 V, the ESDS must be at least 2.5 cm away from the insulator.

Figure 19 illustrates the measurement of an electrostatic potential of 1.4 kV on the surface of a plastic housing at the location where the ESDS is to be installed.

Figure 18.
Electrostatic field intensity measurement on PCB with insulator and without.

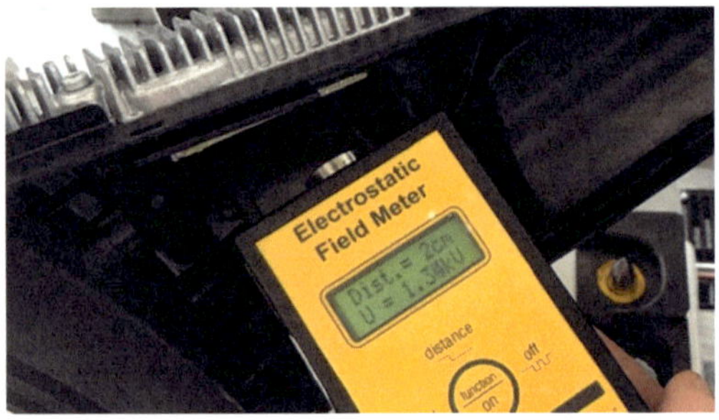

Figure 19.
Electrostatic potential measurement on plastic housing.

4. Case study—Approximation of ESD protection on the automotive headlight production line

The aim of the following part is to propose ESD protection for the production line based on the assumed ESDS threat sources. The case study considers a production line intended for the completion of automotive headlights at the time of its design.

4.1 Production line model

The production line in this study is based on the fact that it ensures the assembly of automotive headlights into their final form. At the entrance to the workplaces of the production line, individual components are delivered, transported, and stored in accordance with the established requirements of ESD protection standards and the internal control program. At the output of the production line, there is a finished lamp (product), which is then packed and transported to the automobile plant. The headlight is then connected there and mounted in the car.

Note: The data presented in this section are created for the purpose of an example of a possible production line concept and do not correspond to any specific production line. The model is created only for the purpose of a better understanding of the issue. Possible similarities with an existing project are purely coincidental.

4.2 Specification of the production line model

The important features (relevant from the point of view of ESD risk) of the compared production lines can be assumed and characterized as follows:

- The production line is in a factory where an ESD control program is established and maintained in accordance with the requirements of IEC 61340-5-1 or ANSI/ESD S20.20.
- The production line completes an automotive headlight with LED technology into which ESDS enters (**Figure 19**).
- Individual components that enter the production process of the production line come ready-made from suppliers or from pre-production workplaces.

- Sensitive ESDS components are packaged and transported in accordance with the requirements of IEC 61340-5-3 (CDM) before delivery to the production line.
- The headlight contains many highly chargeable necessary insulators (typically housing, light guide, lens, etc.), generating and accumulating an electrostatic field.
- The production line consists of manual workplaces where the operations are performed by the operator (HBM) and automatic (robotic) workplaces where the operations are performed without the direct intervention of the operator (IC/MM).
- The final product is in the form of a closed lamp, which is considered by the customer to be a robust component from the point of view of ESD risk (**Figure 20**).

4.3 Description of production line workplaces

Figure 21 illustrates an example of a manual workplace (left) and an automatic workplace (right). In our production line model, these workplaces correspond to positions WP2 and AWS3 (see **Table 2**). The picture also describes the mentioned ESD protection items.

Table 2 describes the workplaces of the production line model and the operations performed. Based on the production operations, it is possible to identify the ESD risk, perform the necessary measurements, and set the ESD protection.

4.4 Identification of ESD threats

Based on the study and description of ESD operations at individual workplaces of the production line model, it is possible to identify ESD threat sources. **Table 3** describes these resources based on operations. These sources can be identified (assumed) from the description, but the ESD measures must be confirmed based on the measurements made (**Table 4**). Only after the measurements have been carried out can the ESD measures be declared valid.

Based on the identified sources of danger in individual operations, it is possible to propose ESD protective steps for workstations. **Table 4** identifies ESD protective measures and suggests measurements to be taken after their implementation.

Tables 2–4 illustrate the design method of ESD protection measures for the automotive headlights production line at the time of its design. This procedure is based on many years of engineering practice and can be expanded or narrowed to the necessary scope according to the workplace or the task being solved.

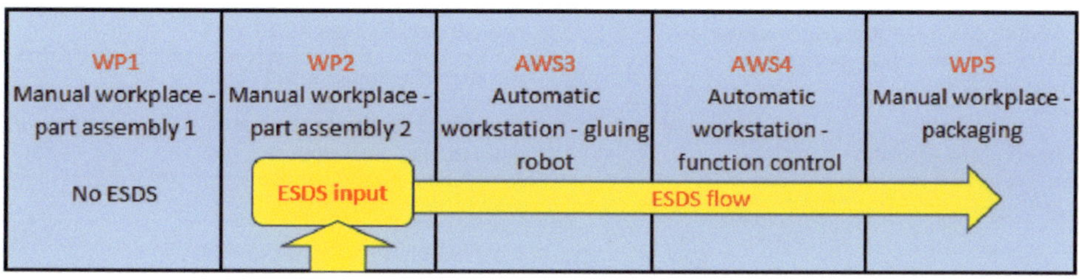

Figure 20.
ESDS flow through the production line.

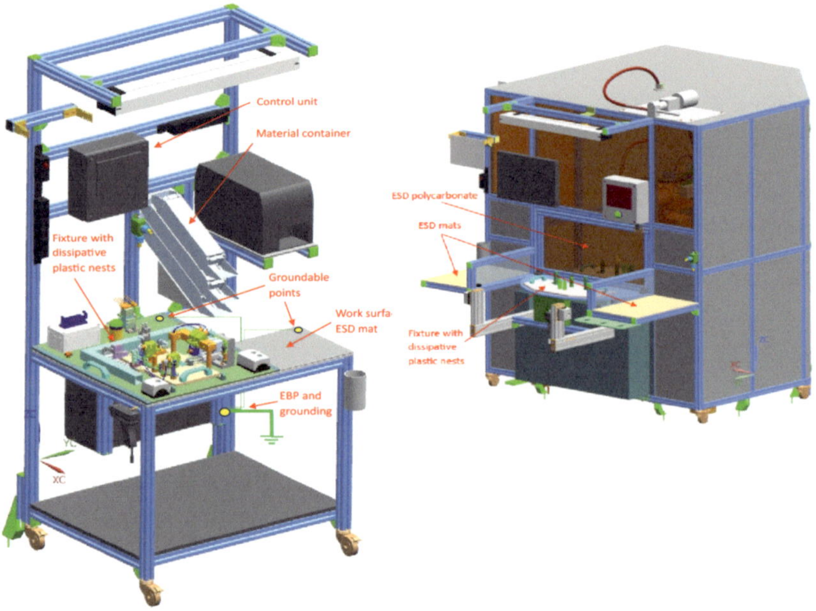

Figure 21.
Example of manual and automatic workplaces with item descriptions.

Workplace	Type	Description of the operation
WP1	Manual workplace—part assembly 1	loading plastic housing into fixture
		installation of cable harness in housing
		installation of plastic components in the housing
WP2	Manual workplace—part assembly 2	manual transfer from WP1
		heatsink (IC) installation
		installation of the PCB light module (ESDS) into housing
		connecting of the PCB light module (ESDS) with heatsink (IC)
		connecting the cable harness to the PCB light module (ESDS)
		light guide and reflector installation
AWS3	Automatic workstation—gluing robot	manual transfer from WP2
		automatic gluing and glass cover installation
AWS4	Automatic workstation—function control	manual transfer from AWS3
		automatic function control - lighting
		automatic function control - measurement and evaluation of quantities
WP5	Manual workplace—packaging	final visual control
		packaging

Table 2.
Description of production line model.

Workplace	Identification of ESD treat sources in operations	Anticipated treats of	ESD model
WP1	essential insulators, operator handling	insulator charging—electrostatic field	none
	essential insulators, operator handling	insulator charging—electrostatic field	
	essential insulators, operator handling	insulator charging—electrostatic field	
WP2	operator handling	insulator charging—electrostatic field	none
	essential insulators, isolated conductor, operator handling	charge of an isolated conductor in an electrostatic field in a housing	IC
	ESDS operator handling	discharge of human body to ESDS during handling	HBM
	essential insulators, isolated conductor, operator handling	charge of ESDS in an electrostatic field, discharge of IC to ESDS, discharge ESDS through the human body	CDM IC HBM
	essential insulators, ESDS operator handling	charge of ESDS in an electrostatic field, discharge ESDS through the human body	CDM HBM
	essential insulators, operator handling	charge of ESDS in an electrostatic field, discharge ESDS through the human body	CDM HBM
AWS3	operator handling	charge of ESDS in an electrostatic field, discharge ESDS through the human body	HBM CDM
	automatic machine operation	charge of ESDS in an electrostatic field, ESD through grounded metal parts	CDM
AWS4	operator handling	charge of ESDS in an electrostatic field, discharge ESDS through the human body	HBM CDM
	automatic machine operation	charge of ESDS in an electrostatic field, ESD through grounded metal parts	CDM
	automatic machine operation	charge of ESDS in an electrostatic field, ESD through grounded metal parts	CDM
WP5	operator handling	charge of ESDS in an electrostatic field, ESD through the human body	HBM
	operator handling	charge of ESDS in an electrostatic field, ESD through the human body	HBM

Table 3.
Identification of ESD threat sources at workstations.

After the assembly of the production line (as part of the trial run), it is necessary to carry out measurements within the entire ESDS flow. Measurements must confirm that the designed ESD protection ensures the entire production process without charging and thus eliminates the risks of ESDS damage. In practice, a situation may arise where the designer of the production line at the time of its design does not know the principles of ESD protection in depth, which can be the cause of incorrect design in the form of under-dimensioning or over-dimensioning of ESD protection measures.

Workplace	Proposed ESD protective measure	Measurement	Note
WP1	not necessary	not necessary	ESDS does not enter the workplace, and insulators are at a sufficient distance from ESDS in the next workplace.
WP2	• grounding a person through ESD footwear and the floor (additionally ESD wrist strap), • ionization to eliminate the charge of essential insulators, • ESD clothing, gloves or finger covers in ESD design, • grounding of surfaces and production components, • plastic nests on a fixture made of dissipative material to eliminate the charge of essential insulators.	personal grounding (Rg/ Rgp, WT) insulators (voltage, field) ESD items (Rg, Rgp, Rpp)	Any object or person that comes into contact with the ESDS must be at the same potential (grounded) with it. At the workplace, the charge must be eliminated by ionization, and, at the same time, recharging must be prevented. Therefore, if there are plastic nests on the fixture, they must be made of dissipative or conductive material.
AWS3	• grounding a person through ESD footwear and the floor, • ionization to eliminate the charge of essential insulators, • ESD clothing, gloves or finger covers in ESD design, • grounding of surfaces and production components in machine, • plastic nests on a fixture made of dissipative material to eliminate the charge of essential insulators.	personal grounding (Rg/ Rgp, WT) insulators (voltage, field) ESD items (Rg, Rgp, Rpp)	Any object or person that comes into contact with the ESDS must be at the same potential (grounded) with it. At the workplace, the charge must be eliminated by ionization, and, at the same time, recharging must be prevented. Therefore, if there are plastic nests on the fixture, they must be made of dissipative or conductive material. Ionization in the machine is necessary if there is charging in the process.
AWS4	• grounding a person through ESD footwear and the floor, • ionization may be necessary if charging occurs in the process, • ESD clothing, gloves or finger covers in ESD design, • grounding of surfaces and production components in machine, • plastic nests on a fixture made of dissipative material to eliminate the charge of essential insulators.	personal grounding (Rg/ Rgp, WT) insulators (voltage, field) ESD items (Rg, Rgp, Rpp)	Any object or person that comes into contact with the ESDS must be at the same potential (grounded) with it. At the workplace, the charge must be eliminated by ionization, and, at the same time, recharging must be prevented. Therefore, if there are plastic nests on the fixture, they must be made of dissipative or conductive material. Ionization in the machine is necessary if there is charging in the process.
WP5	• grounding a person through ESD footwear and the floor, • ESD clothing, gloves or finger covers in ESD design, • grounding of surfaces and production components in machine.	personal grounding (Rg/ Rgp, WT) insulators (voltage, field) ESD items (Rg, Rgp, Rpp)	On the final WP, it is assumed that the automotive headlight is closed and the risk of ESD damage is low. It is necessary to take into account the type of packaging and the customer's requirements.

Table 4.
Approximation of ESD protective measures in production line model.

One of the ways of adapting ESD protection measures to a specific situation is the analysis and evaluation of ESD risks. Documents such as IEC PAS 61340-5-6 and ANSI/ESD SP17.1 provide methods and technology for assessing processes from the point of view of ESD risks. These standards contain a general approach to assessing ESD risks using various measurement methods [3, 12]. Thus, the implementation of ESD protection items is necessarily related to the anticipated risk of ESDS damage.

5. Conclusion

LED technology in automotive headlights is susceptible to damage from electrostatic discharge (ESD), requiring protective measures per IEC 61340-5-1 or ANSI/ESD S20.201 standards. ESD protection includes administrative and technical measures like grounding, personal grounding, workstation protection, and packaging. This article outlines common measurement methods for identifying ESD in manufacturing, focusing on verifying grounding requirements for personnel and ESD protection items, including measurements on necessary insulators.

The case study focuses on a typical production line for automotive headlights, analyzing the processes and probability of occurrence of electrostatic potential. It describes risky situations in production lines where electrostatic discharges can occur and suggests protective measures to eliminate these risks.

Based on the given information, it is possible to create your own procedures for setting measures, but a detailed study and understanding of the theory is necessary. Within widely available resources, ESD protection is discussed mainly in relation to ESDS during their production. However, a less resolved issue is the implementation of ESD protection in production lines in such a way as to enhance efficiency and cost-effectiveness.

Practice shows that risk mitigation measures are often insufficient in high-risk workplaces and over-engineered in low-risk ones. This issue is commonly identified during trial operations. A well-prepared ESD risk analysis can simplify the designer's work, optimize resources and time in designing ESD measures, and improve economic efficiency. Properly designing ESD protection from the start allows for better cost estimation, reduces deviations, and lowers the costs of addressing them [1].

Acknowledgements

AI translation tool DeepL.com was used in the translation process of some parts of this monograph. The intellectual content and analysis are the full creation of the authors.

Author details

Michal Sahul* and Peter Janiga
Institute of Power and Applied Electrical Engineering, Slovak University of Technology in Bratislava, Bratislava, Slovakia

*Address all correspondence to: sahul@e-forrest.sk

References

[1] IEC 61340-5-1 (2024-05) Ed. 3.0: "Electrostatics—Part 5-1: Protection of Electronic Devices from Electrostatic Phenomena—General Requirements". Geneva, Switzerland: IEC Technical Committee 101, International Electrotechnical Commission (IEC); 2024

[2] Tamminen P, Viheriäkoski T, Ukkonen L, Sydänheimo L. ESD and disturbance cases in electrostatic protected areas. In: 2015 37th Electrical Overstress/Electrostatic Discharge Symposium (EOS/ESD), Reno, NV, USA: IEEE; 2015. pp. 1-7. DOI: 10.1109/EOSESD.2015.7314792

[3] Zhou P, Leung HKN. An integrated risk analysis method using spatial interpolation. In: 2012 19th Asia-Pacific Software Engineering Conference, Hong Kong, China: IEEE; 2012. pp. 452-461. DOI: 10.1109/APSEC.2012.11

[4] IEC TR 61340-1:2012+AMD1:2020 CSV (2020-06) Ed. 1.1: "Electrostatics—Part 1: Electrostatic Phenomena—Principles and Measurements." Geneva, Switzerland: IEC Technical Committee 101, International Electrotechnical Commission (IEC); 2020

[5] IEC 61340-5-3 (2022-04) Ed. 3.0: "Electrostatics—Part 5-3: Protection of Electronic Devices from Electrostatic Phenomena—Properties and Requirements Classification for Packaging Intended for Electrostatic Discharge Sensitive Devices". Geneva, Switzerland: IEC Technical Committee 101, International Electrotechnical Commission (IEC); 2022

[6] IEC TR 61340-5-2:2018 RLV (2018-03) Ed. 2.0: "Electrostatics—Part 5-2: Protection of Electronic Devices from Electrostatic Phenomena—User Guide". Geneva, Switzerland: IEC Technical Committee 101, International Electrotechnical Commission (IEC); 2018

[7] IEC 60749-26:2018 (2018-01) Ed. 4.0: "Semiconductor Devices—Mechanical and Climatic Test Methods—Part 26: Electrostatic Discharge (ESD) Sensitivity Testing—Human Body Model (HBM)". Geneva, Switzerland: IEC Technical Committee 47, International Electrotechnical Commission (IEC); 2018

[8] Sahul M, Janiga P. Optimization of ESD protection design for assembly lines of headlights through risk analysis. In: 2024 24th International Scientific Conference on Electric Power Engineering (EPE). Ostrava Czech Republic: VSB – Technical University of Ostrava; 2024. pp. 1-5. DOI: 10.1109/EPE61521.2024.10559564

[9] ESD TR53-01-2018 Compliance Verification of ESD Protective Equipment and Materials, Electrostatic Discharge Association. Rome, NY, USA: ESDA Standards Committee, Electrostatic Discharge Association (ESDA); 2018

[10] IEC 61340-4-5:2018 RLV (2018-01) Ed. 2.0: "Electrostatics—Part 4-5: Standard Test Methods for Specific Applications—Methods for Characterizing the Electrostatic Protection of Footwear and Flooring in Combination with a Person". Geneva, Switzerland: IEC Technical Committee 101, International Electrotechnical Commission (IEC); 2018

[11] IEC 61340-2-3:2016 RLV (2016-06) Ed. 2.0: "Electrostatics—Part 2-3:

Methods of Test for Determining the Resistance and Resistivity of Solid Materials Used to Avoid Electrostatic Charge Accumulation". Geneva, Switzerland: IEC Technical Committee 101, International Electrotechnical Commission (IEC); 2016

[12] Standard, IEC PAS 61340-5-6:2022 Electrostatics—Part 5-6: Protection of Electronic Devices from Electrostatic Phenomena—Process Assessment Techniques. Geneva, Switzerland: IEC Technical Committee 101, International Electrotechnical Commission (IEC); 2022

Chapter 4

Electromagnetic Sensors

Xing Fan, Zhehao Pei and Qingrong Peng

Abstract

This chapter introduces various types of sensors used to detect electromagnetic fields, detailing their mechanisms and applications. The discussed sensors include widely used electric and magnetic field sensors based on electromagnetic induction principles, rapidly advancing sensors based on optical effects, and newly proposed quantum-based sensors. A significant focus is placed on sensors utilizing optical effects, which have seen rapid development and practical application thanks to advancements in laser and crystal material technologies. Recent findings from our research team are also presented. Optical effect sensors, characterized by their compact size and full insulation, are particularly suitable for measuring electromagnetic fields in harsh electromagnetic environments and confined spaces. Meanwhile, quantum effect sensors offer extremely high sensitivity and can perform absolute electric field measurements, making them promising for applications such as electric field calibration and monitoring of extremely weak electric fields.

Keywords: electromagnetic sensor, electric sensor, optical-electro sensor, quantum effect, electromagnetic field measurement

1. Introduction

The final chapter of this work focuses on the measurement of electric and magnetic field parameters, aiming to provide a comprehensive overview of various sensors used for this purpose. The measurement of electric and magnetic fields fundamentally relies on the interaction between materials and electromagnetic fields, whether these materials are metallic or non-metallic. Consequently, any measurement process inherently involves some level of interaction, which can disturb the measurement itself. This disturbance is a common feature of all observational methods. Thus, understanding the measurement principles and their impact on the measured parameters is crucial for selecting the optimal sensor.

This chapter introduces several types of sensors, including those based on electromagnetic induction principles, optical effects, and recent advancements in quantum-based measurements. The goal is to offer readers a thorough understanding of electromagnetic measurement techniques and to promote their application across various scientific and engineering fields.

1.1 Section 1: Electromagnetic induction-based sensors

Electromagnetic induction sensors represent some of the earliest, most mature, and widely used methods for measuring electric and magnetic fields. For electric field measurements, these sensors utilize a capacitive induction mechanism involving a dual-electrode structure. The change in induced charge between two electrode plates, under the influence of an external electric field, is proportional to the field strength. The electric field strength is then inferred from the induced current measured between these plates.

For magnetic field measurements, electromagnetic induction is employed through a coil structure. The voltage generated across the coil in the presence of an external magnetic field is proportional to the magnetic flux through the coil, allowing the magnetic field strength to be determined from the measured induced voltage. These sensors are characterized by their simple structure and moderate sensitivity. However, they have drawbacks such as large volume and high metal content, which can affect the measured electromagnetic field. They are suitable for open-space measurements, such as detecting electromagnetic environments around transmission lines, measuring lightning overvoltages, and assessing radiation fields around communication stations.

1.2 Section 2: Optical effect-based sensors

The second section delves into sensors based on optical effects, which emerged in the latter half of the twentieth century. With advances in laser and crystal technologies, these sensors have evolved rapidly and become increasingly practical. These sensors exploit nonlinear optical effects in optical crystals, including the Pockels effect in electro-optic crystals, the Faraday rotation effect in magneto-optical crystals, and the Kerr effect in certain materials.

In electric field measurements, the Pockels effect is commonly used. This effect relies on the fact that the refractive index of electro-optic crystals changes proportionally with the applied electric field. The Kerr effect, on the other hand, is proportional to the square of the refractive index and the electric field strength. For magnetic field measurements, the Faraday rotation effect is employed. This effect causes a differential refractive index for left- and right-handed polarized light under the influence of a magnetic field, resulting in a rotation of the light's polarization direction. The angle of this rotation, known as the Faraday rotation angle, is proportional to the magnetic field strength.

Optical effect-based sensors offer advantages such as complete insulation, wide response frequency bands, miniaturization, and strong anti-interference capabilities. However, their performance can be significantly influenced by the stability of the laser system. These sensors are particularly suitable for measuring electromagnetic fields in harsh environments and confined spaces, and they have found applications in monitoring large power equipment like transformers and insulated switchgear, demonstrating considerable potential.

1.3 Section 3: Quantum effect-based sensors

The final section covers quantum effect-based sensors, a recent development in the field. Despite their nascent stage, these sensors have shown considerable performance advantages. They primarily utilize the electromagnetically induced transparency (EIT) effect. In this phenomenon, two light beams are directed simultaneously onto an atomic medium; one beam resonates with an atomic transition, passing through the medium without absorption or reflection. When an external electric or magnetic field is applied, the EIT spectrum's main peak splits into two, a phenomenon known as electromagnetically induced transparency autler-townes splitting (EIT-AT). The frequency difference between these two transmission peaks is used to calculate the external field's strength.

Quantum effect-based sensors are noted for their exceptionally high sensitivity, which is more than three orders of magnitude greater than that of traditional electromagnetic induction and optical effect-based sensors. This high sensitivity suggests significant potential for future development and application.

In summary, this chapter presents a detailed exploration of various electromagnetic field sensors, providing insights into their principles, advantages, and applications. Each sensor type offers unique benefits and challenges, making them suitable for different measurement contexts and requirements.

2. Electric sensors

2.1 Conventional magnetic sensor

2.1.1 Fluxgate sensor

The fluxgate sensor measures the magnetic field through the nonlinear relationship between the magnetic induction intensity of a magnetic core and the external magnetic field's intensity under alternating magnetic field saturation. This sensor consists of an iron core wound with two coils: the main (excitation) coil and the auxiliary (collection) coil [1]. During operation, an excitation current is applied to the main coil to saturate the iron core. When unsaturated, the core provides a low magnetic resistance path due to its higher permeability than the external field. When saturated, the magnetic resistance increases, causing magnetic lines of force to overflow the core. The external magnetic field is then extracted from the output signal using principles such as the second harmonic, pulse positioning, or pulse height methods.

Fluxgate sensors are primarily used for measuring weak, constant, or slowly varying magnetic fields and are valued for their high resolution and broad measurement range.

2.1.2 Hall effect sensors

The Hall effect describes the generation of an electromotive force when a magnetic field perpendicular to the current in a metal or semiconductor creates a potential difference. This potential difference is directly proportional to the current and magnetic field intensity and inversely proportional to the plate thickness [2]:

$$U_H = R_H \frac{IB_0}{d} \tag{1}$$

where R_H is the material-related Hall coefficient; d is the thickness of the Hall device; I is the current flowing through the Hall device; U_H is the magnetic induction intensity of the external magnetic field.

When the current is constant, it is proportional to the external magnetic field, and the proportionality coefficient is [3, 4]:

$$K = R_H \frac{I}{d} \tag{2}$$

If the proportional coefficient of the Hall plate K is known and the I and U_H is measured by an instrument, the magnetic induction intensity can be calculated.

The Hall sensor has a large magnetic field measurement range and is widely used to measure various components of alternating current (AC), DC magnetic fields and pulse magnetic fields. It can also be used to measure gap magnetic fields.

2.1.3 Magnetoresistive sensor

The magnetoresistive sensor functions based on the magnetoresistive effect in anisotropic magnetic materials. When current flows through the material, its resistance varies with the angle between the current and magnetization directions. An applied magnetic field causes the magnetization direction to rotate. If it turns perpendicular to the current, the resistance decreases. If parallel, the resistance increases. Typically, four such resistors form a bridge circuit in the sensor. When a magnetic field is applied, the resistance of two opposite resistors increases, while the other two decrease. Within its linear range, the bridge's output voltage is proportional to the measured magnetic field [5].

Currently, the magnetoresistive effect is widely used in magnetic sensing, magnetometers, electronic compasses, position and angle sensors, and other fields.

2.2 Conventional electric field sensors

2.2.1 Resistance splitter

A resistor shunt is an electronic component that realizes current distribution by connecting resistors of different resistance values in series. Its working principle is based on Kirchhoff's law and Ohm's law. It shows high accuracy and fast response speed when detecting low-frequency and small-amplitude currents.

2.2.2 Current transformer

Current transformers use electromagnetic induction principles to convert high currents into lower secondary currents for measurement and protection purposes. They are commonly used in power systems for current measurement, fault protection, and energy metering.

2.2.3 Hall sensor

Hall sensors measure current based on the Hall effect, determining current magnitude from the Hall potential induced by the magnetic field. They are applicable for AC and DC electric field measurements and provide high accuracy and isolation.

2.2.4 Giant magnetoresistive current sensor

The giant magnetoresistance sensor is a sensor that uses the giant magnetoresistance effect to measure the magnetic field. It usually consists of four giant magnetoresistances forming a Wheatstone bridge structure, two active resistors and two passive resistors. When an external magnetic field acts on the giant magnetoresistance sensor, the resistance value in the bridge will change slightly, causing the bridge to be unbalanced. This unbalanced signal can be converted into the measured value of the electric field after passing through the signal processing circuit.

With the Wheatstone bridge structure, the sensor can realize high-sensitivity magnetic field measurement, which is widely used in AC and DC electric field measurement [6–11].

2.2.5 Rogowski coil

The Rogowski coil, or magnetic potential meter, operates on electromagnetic induction. An AC current through the coil generates a changing magnetic field that induces an electromotive force. Measuring this induced voltage allows accurate inference of the current in the original circuit. Rogowski coils are ideal for strong AC electric field measurements.

3. Optical-electro sensors

3.1 Optical magnetic field sensor

3.1.1 Magnetic field sensor based on magnetostrictive effect

When a ferromagnetic body in an external magnetic field is magnetized, its length and volume will change. This phenomenon is called magnetostrictive effect. The combination of fiber Bragg grating (FBG) and magnetostrictive effect can realize the measurement of magnetic field. The principle of FBG magnetic field sensing is to stick the fiber Bragg grating on the magnetostrictive material. The light emitted by the broadband light source is incident from one end of the FBG. Due to the periodic change of the refractive index, the light waves transmitted forward and backward in the fiber core are coupled. When the Bragg condition is met, that is:

$$\lambda_B = 2n\Lambda \tag{3}$$

where λ_B is the central wavelength of the incident light (Bragg wavelength), n is the effective refractive index of the core in the grating region, and Λ is the FBG grating period.

The effect of magnetic field on magnetostrictive material can be used to modulate fiber Bragg grating, and the magnitude of magnetic field can be determined by measuring the Bragg wavelength drift.

3.1.2 Magneto-optical crystal magnetic field sensor

When polarized light traverses certain anisotropic media under the influence of a magnetic field, it induces alterations in the electromagnetic properties of the medium, resulting in the rotation of the plane of polarization (the plane in which the electric field vibrates) of the light. This phenomenon is known as the magneto-optical effect. The magneto-optical effect method is an approach for measuring magnetic fields by utilizing the magneto-optical effect produced through the interplay between the magnetic field, light, and medium. By employing magneto-optical crystals characterized by a high Verdet constant as the sensing element, the magnitude of the magnetic field under investigation can be determined from the deflection angle of the polarized light.

3.1.3 All-fiber magnetic field sensor

The sensing part of an all-fiber-optic optical sensor consists of a winding of sensitive optical fibers around a current-carrying conductor, the sensitivity of which is determined by the length of the sensing fiber [12–14]. The linearly polarized light passes through the $\lambda/4$ slide and becomes circularly polarized. The circularly polarized light arriving at the reflector is reflected by the reflector and returns in the same way, doubling the angle of deflection due to the non-reciprocity. The photodetector receives the reflected signals and processes them to obtain the magnetic field and current signals.

3.2 Optical electric field sensor

3.2.1 Inverse piezoelectric effect sensor

The inverse piezoelectric effect describes the phenomenon whereby applying an electric field to a piezoelectric material induces mechanical deformation or stress in a specified direction. Upon removal of the electrical field, the deformation ceases to exist [15, 16]. This effect can be exploited in electric field sensors by translating the minute deformations of the crystal into optical signals. These signals are subsequently modulated and detected to facilitate precise measurements of the electric field or voltage.

An optical electric field sensor utilizing the inverse piezoelectric effect offers a notable advantage over systems that require electro-optical crystals: it inherently reduces interference from extraneous optical effects on the sensing signal, thereby providing a cost-effective solution.

3.2.2 Electric field sensor based on magneto-optical effect

When polarized light traverses certain anisotropic media in the presence of a magnetic field, it results in modifications to the medium's electromagnetic properties and induces a rotation in the light's plane of polarization (i.e., the plane of electric field vibration). This phenomenon is known as the magneto-optical effect. The magneto-optical effect method is a technique employed to measure an electric field by leveraging the magneto-optical effect, which arises from the interaction between the magnetic field—generated by the electric field under examination—and the light and the medium. By utilizing a magneto-optical crystal with a high Verdet constant as a sensing unit, the intensity of the electric field can be determined through the deflection angle of the polarized light.

3.2.3 Kerr effect electric field sensor

The Kerr effect refers to the phenomenon where the refractive index of an electro-optical crystal is proportional to the square of the applied electric field's magnitude. The relationship between the change in the crystal's refractive index and the applied electric field is given by Refs. [9, 17–20]:

$$\Delta n = KE^2 \tag{4}$$

In the formula, Δn represents the change in the refractive index of the electro-optic crystal, K is a constant related to the crystal and light wavelength, E is the electric field intensity applied to the crystal, and Δn is the change in the polarization state of the light passing through the crystal. The measured electric field is determined by detecting the polarization state of the light wave.

The Kerr effect is very weak and not linearly related to the applied electric field, making the signal challenging to demodulate.

3.3 Temperature characteristics

Based on the foundational principles of electro-optic sensing, we have developed an experimental platform designed to measure the working point and phase shift of an electro-optic (EO) crystal sensing system subjected to combined temperature and electric field influences, as depicted in **Figure 1** [21–27]. The experimental setup comprises a polarization state modulation optical system centered around the EO crystal, a consistent electric field generation system using a flat plate electrode, and a temperature control and monitoring setup based on a semiconductor thermo-electric chip. The optical system for polarization state modulation consists primarily of a narrow linewidth semiconductor laser with a wavelength of 1550 nm, a polarization beam splitter (PBS) prism with an extinction ratio of 1000:1, a quarter-wave zero-order cemented wave plate, an EO crystal, and an indium gallium arsenide photodetector operative within the range of 800 to 1700 nm. A high voltage at utility frequency is applied to the flat plate electrode, creating an electric field that varies from 0 to 1000 kV/m across the gap. The temperature control system integrates a semiconductor thermo-electric chip, a PT100 thermistor, a temperature controller, a current-type drive board, and a water medium heat exchange system. This configuration allows for maintaining the temperature within a range of −20 to 60°C, with a precision of 0.1°C.

Figure 2 presents a physical depiction of the experimental platform. At the core is the optical path system for polarization state modulation, utilizing electro-optic crystals. Parallel metal plates are linked to high-voltage electrodes, with the crystal placed on a metal plate that is integrated into the temperature control system. Temperature

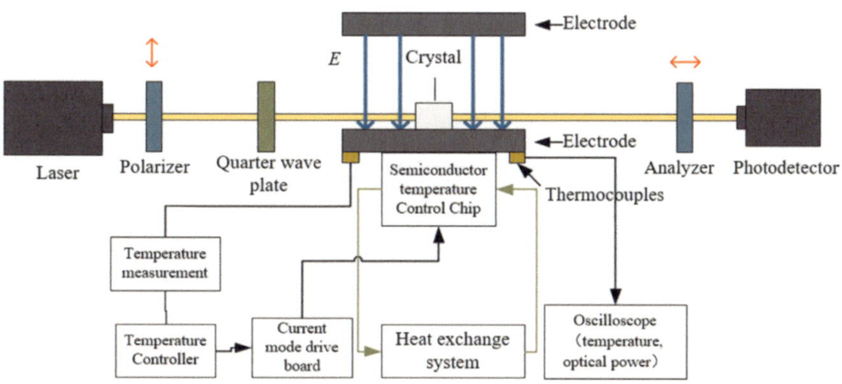

Figure 1.
Schematic diagram of the electro-optic sensor temperature characteristic experiment platform.

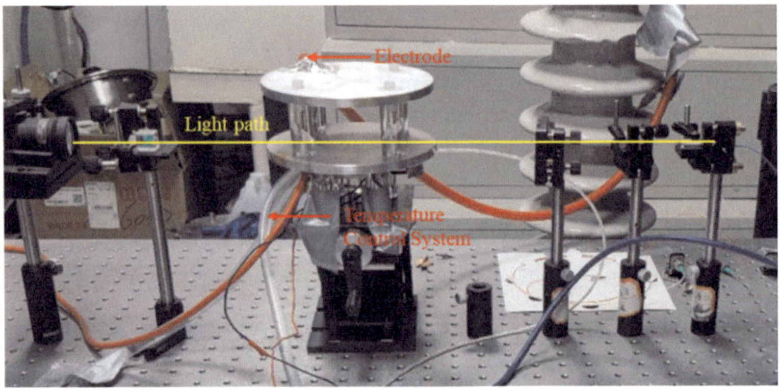

Figure 2.
Physical illustration of the experimental platform.

regulation is accomplished through thermal equilibrium between the crystal and the plate electrode.

The EO crystal is a crucial part of the electric field sensor, with its optical properties impacting sensitivity, measurement range, accuracy, response speed, and other sensor attributes. The chosen crystal should possess stable properties, a high EO coefficient, excellent optical uniformity, and good light transmittance. After a thorough evaluation, we selected BGO crystal and z-axis light-transmitting lithium niobate (LN) as the EO crystals for the sensor.

The thermal characteristics of the electro-optic (EO) sensor output were evaluated using a rigorously designed experimental setup. The EO crystal was situated between electrodes, and once the platform's temperature stabilized at the specified level, we measured the sensor's operating point and modulation signal at various voltages. Each data point was measured 30 times to ensure accuracy, with the mean value calculated and a measurement error maintained within 0.5%. Data for the operating point and modulation signal were recorded for the bismuth germanate (BGO) sensor over the temperature range of 20–30°C and for the lithium niobate (LN) sensor over 20–34°C.

As illustrated in **Figures 3** and **4**, the operating point and modulation signal for both the BGO and LN sensors show periodic and synchronous fluctuations with increasing temperatures. Notably, the temperature periods differ significantly between the two sensors. The BGO sensor exhibits a temperature period of 2°C, while the LN sensor displays a temperature period of 11°C.

To assess the fluctuation amplitude of the operating point and EO (Electro-Optic) modulation signal concerning temperature, we define the fluctuation amplitude, denoted as γ, as the ratio of the difference between the maximum and minimum values of the signal to the maximum light intensity. This parameter serves as an indicator of the sensor's temperature stability, with a smaller γ value signifying enhanced temperature stability:

$$\gamma = \frac{I_{\max} - I_{\min}}{I_{\max}} \tag{5}$$

The experimental data demonstrate that the BGO sensor's working point exhibits a fluctuation amplitude of 0.38 and a modulation signal amplitude of 0.60, corresponding with a temperature cycle of 2°C. In contrast, for the LN sensor, both the working point and the modulation signal present a fluctuation amplitude of 0.42,

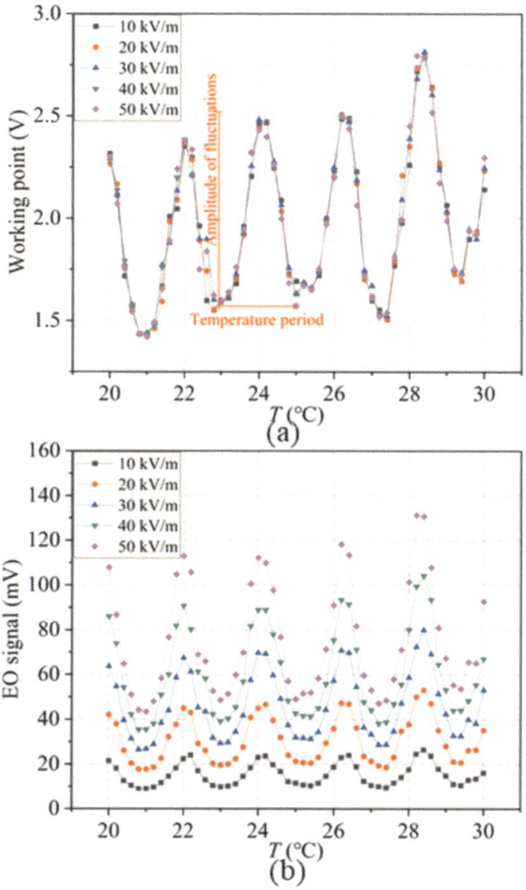

Figure 3.
BGO sensor's temperature characteristics: (a) working point and (b) EO signal.

associated with a temperature cycle of 11°C. As depicted in **Figure 5**, despite variations in temperature, the modulation signals of both electro-optic sensors maintain a robust linear correlation with the applied electric field, although the slope of this correlation is affected by temperature.

3.3.1 Model of a crystal resonator

The BGO in the experiment is an isotropic crystal, and the refractive indices of the x and y axes of LN are identical, removing the impact of natural birefringence on the sensor. However, the sensor's output continues to undergo periodic fluctuations with temperature variations.

3.3.2 Analysis of the mechanism

In **Figure 6**, light propagates through the crystal while undergoing continuous processes of reflection and transmission. The portion of light that avoids reflection at the inner surface of the crystal is referred to as zero-order transmitted light, whereas light experiencing m reflections and transmissions is classified as mth-order

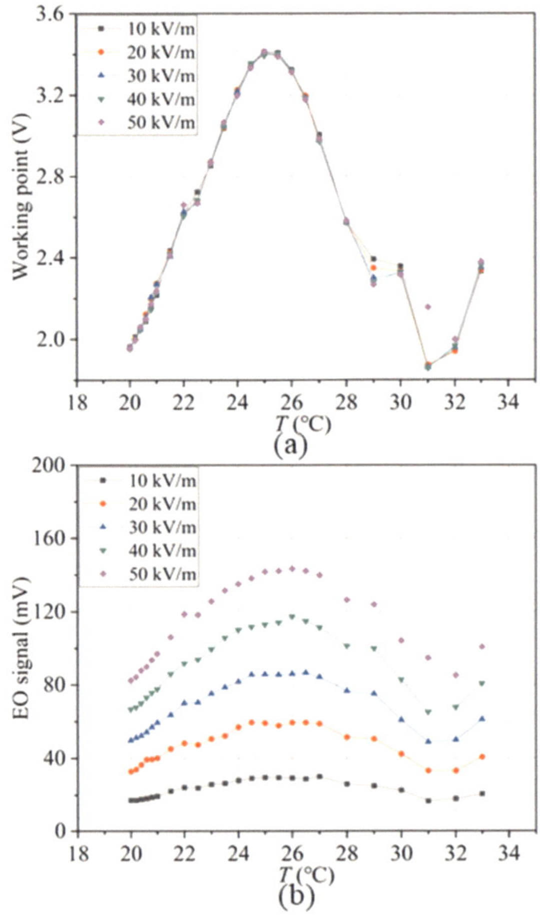

Figure 4.
Temperature characteristics of LN sensor: (a) working point and (b) EO signal.

transmitted light. The light eventually emerging from the crystal results from the coherent superposition of multiple transmission orders produced by the resonant cavity. This interference pattern is highly sensitive to variations in the phase difference among the transmission orders.

As depicted in **Figure 7**, the refractive index and dimensions of the crystal are modified due to its thermo-optic effect and thermal expansion when subjected to strong interactions between temperature and electric fields. These changes alter the optical thickness of the crystal resonator cavity, thereby affecting the phase difference and, consequently, the interference intensity. The periodic nature of the influence on the phase difference accounts for the periodic fluctuations observed in thermal-induced sensor responses.

3.3.3 Sensor response under the AC electric field and DC electric field

The output characteristics of the EO sensor were analyzed under both alternating current (AC) and direct current (DC) electric fields, revealing substantial differences between the two conditions, as depicted in **Figure 8**.

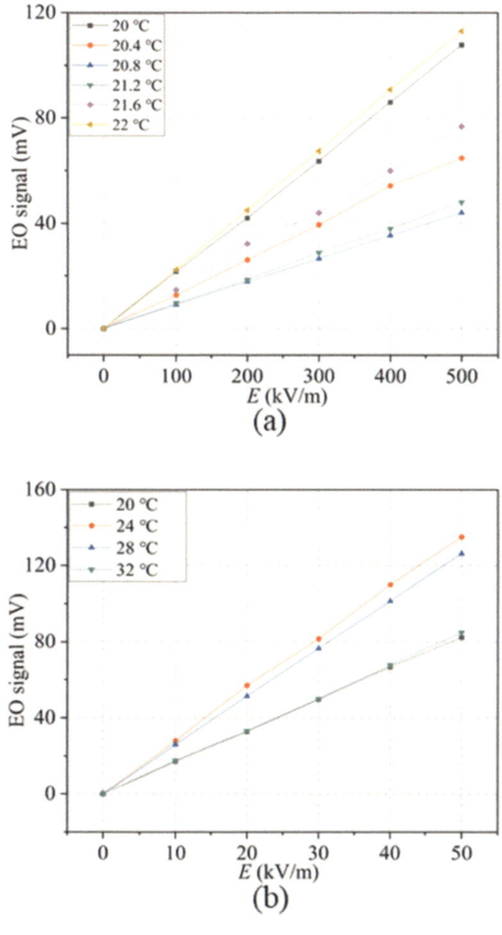

Figure 5.
The relationship between the EO signal and the electric field. (a) BGO sensor and (b) LN sensor.

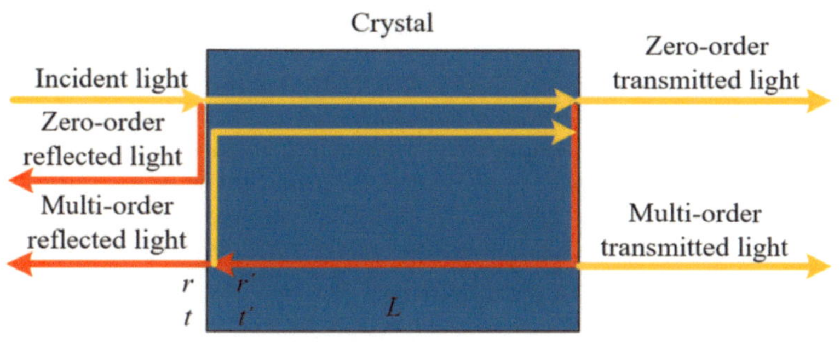

Figure 6.
Model of the electro-optic crystal resonator. (r *and* t *indicate the reflection and transmission values at the external surface,* r' *and* t' *represent the reflection and transmission coefficients at the inner interface, and* L *is the length of the crystals.)*

To align the experimental data, the least squares fitting method was employed. The results demonstrated that both output curves exhibited exceptional linearity, with coefficients of determination (R^2) of 0.9995 and 0.9997, respectively. The sensitivity

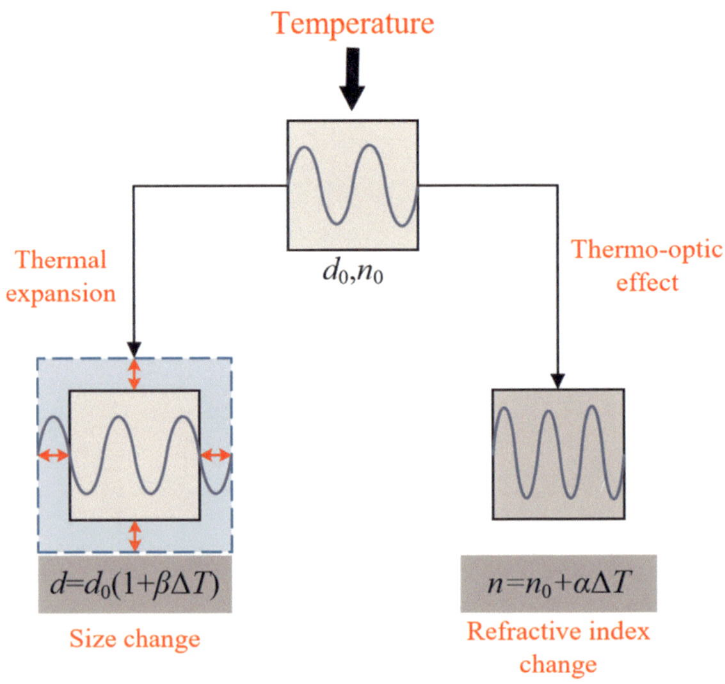

Figure 7.
The effect of temperature on the phase of light beams in crystals.

of the measurement was defined as the change in sensor output per unit change in electric field intensity:

$$S = \frac{dU}{dE} \tag{6}$$

Here, S represents the sensitivity, U denotes the sensor's output, and E stands for the applied electric field. In the linear range, the sensitivity can be considered as the gradient of the output curve.

This study determined that the sensitivity to direct current (DC) electric fields is approximately 0.1 mV/(kV/m), while for alternating current (AC) electric fields, it is around 0.06 mV/(kV/m). Analysis of the residual error graph indicates that when a DC field of 376 kV/m is applied, the maximum residual error in the output voltage signal is approximately 0.5 mV. This corresponds to a measurement error of 4.9 kV/m, which constitutes a relative error of 1.3%. In the case of a 379 kV/m AC field, the residual error in the output voltage is about 0.2 mV, equating to a measurement error of 3.2 kV/m and a relative error of 0.8%. Throughout the entire testing range, the maximum relative error remains below 3%, thus conforming to the national standard's error threshold of 3% for high voltage measurements.

Based on electro-optic (EO) theory and dielectric polarization theory, the measurement sensitivity is primarily influenced by the EO coefficient, the dielectric constant, and the size of the crystal. Previous studies have indicated that various types of electric fields do not exert a significant influence, although these findings were predominantly derived from studies involving AC or high-frequency fields. In the context of this research, it is essential to acknowledge that when measuring DC electric

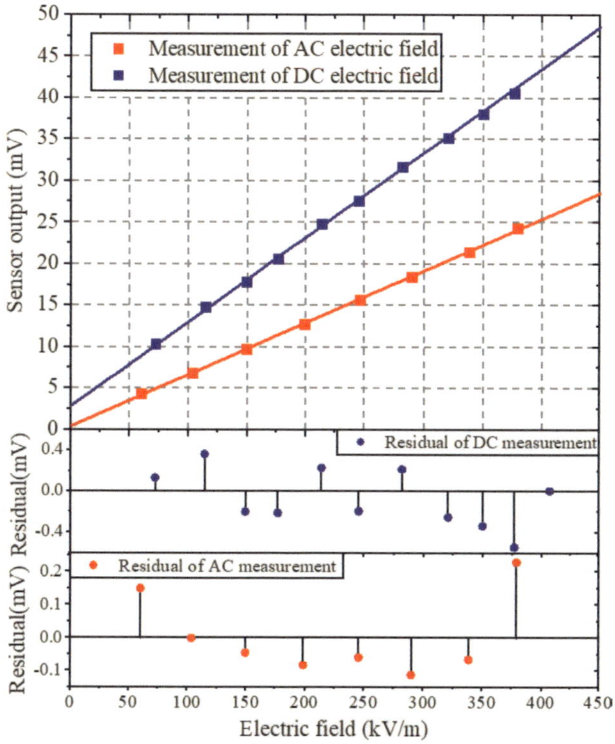

Figure 8.
Comparison of response curves for the EO sensor under AC and DC electric field.

fields, the conductivity of the EO crystal significantly affects its internal field, presenting substantial differences from measurements under AC or high-frequency conditions. According to electrodynamics theory, the distribution of electric fields in dielectrics differs between DC and AC fields. Under an AC field, the electric field distribution within dielectrics conforms to the fundamental and constitutive equations of the electrostatic field:

$$\begin{cases} \nabla \cdot D = \rho \\ \nabla \times E = 0 \\ D = \varepsilon E \end{cases} \tag{7}$$

In a DC electric field, the electric field distribution within dielectrics complies with the fundamental and constitutive equations of the steady current field:

$$\begin{cases} \nabla \cdot J = 0 \\ \nabla \times E = 0 \\ J = \gamma E \end{cases} \tag{8}$$

The distribution of electric fields in dielectrics is predominantly influenced by the dielectric constant in an AC field, while conductivity is the main factor under a DC field. This distinction presents a significant challenge for electro-optic (EO) sensors measuring DC fields in air, as surface charge accumulation caused by the conductive properties of the crystal impairs the internal electric field. In contrast, when immersed in oil, the conductivity is considerably higher, and the resultant charge accumulation due to oil conductivity substantially affects the internal field of the crystal. Due to the interplay of conductance between the oil and the crystal, the accumulated charges at the interface do not entirely cancel out the internal electric field of the EO crystal as they do in an air environment. On the contrary, such charges may enhance the internal field, enabling the sensor to effectively respond to DC fields, as demonstrated by our experimental observations. Consequently, the sensitivity of the EO sensor to DC fields is influenced by both the conductivity of the crystal and the oil, while its sensitivity to AC fields is governed by the dielectric constant. This variance in sensitivities is further explored and confirmed through theoretical analysis and simulations in Section IV.

3.3.4 Response of the sensor to a hybrid AC/DC electric field

The waveform parameters for the AC/DC hybrid electric fields measured by the electro-optic sensor are defined and illustrated in **Figure 9**.

In our study, we define the amplitude of the direct current (DC) component of hybrid electric fields as the value observed at the point of maximum derivative. Conversely, the amplitude of the alternating current (AC) component is defined as half the difference between the maximum and minimum values of the electric fields. It is important to note that harmonic voltages originating from AC sources are excluded from consideration. Generally, the DC component is determined by averaging the data from hybrid electric fields, whereas the AC component is extracted through the application of filtering techniques.

The direct current (DC) output characteristics of the electro-optic (EO) sensor, incorporating varying alternating current (AC) components, are illustrated in **Figure 9**. This figure showcases data set A, which has been computed using the periodic average method. The formula employed for this method is as follows:

$$U_{\text{dc\_average}} = \frac{\sum_{i=1}^{n} U_i}{n} \tag{9}$$

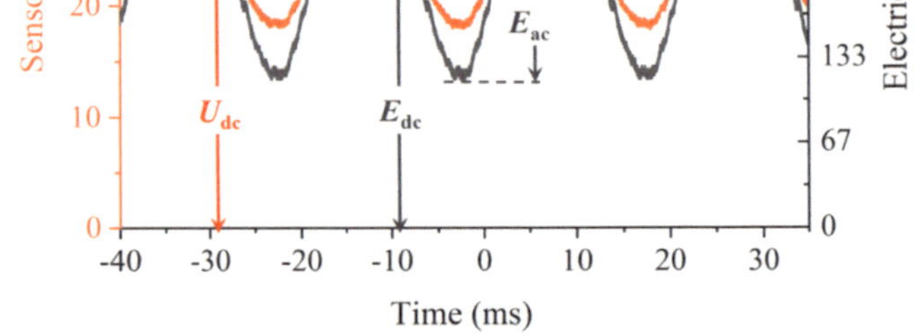

Figure 9.
Waveform of the AC/DC hybrid electric field with defined parameters.

In this context, Udc_average is the computed DC component, Ui represents the voltage at a given sampling point, U1to Un denote the voltage values at sampling points within an AC cycle, and n is the total number of sampling points in that AC cycle.

The DC measurement signal in data set A is affected by the AC electric field. As the AC field intensifies, the slope of the DC sensing curve slightly reduces. This altering effect of the AC field on DC measurement sensitivity has not been previously documented. We believe the EO sensor's optical nonlinearity is responsible for this interaction. The EO modulation's optical function is inherently nonlinear, and the sensor output remains linear only if the condition $E < <E\pi/2$ is met. When the combined amplitude of the hybrid fields nears or surpasses the linear range, the sensor's nonlinear output can lead to errors in the DC signal.

In this paper, $E\pi/2$ of the EO sensor is about 7 MV/m. However, when the AC electric field increases to 1.6 MV/m, the condition of $E < <E\pi/2$ is no longer satisfied. The AC component in the modulation signal PE will be distorted, and asymmetrical distortion part in AC signal will be added to the DC data calculated by the period average method. This affects the AC electric field on the DC output curve. To verify this idea, the maximum-derivative method is utilized to extract the DC component of the sensor output signal. Because this method calculates the amplitude of a DC signal by selecting the zero-crossing point of an AC signal, the influence of asymmetrical distortion of an AC signal can be avoided.

In this study, the EO sensor's $E\pi/2$ is approximately 7 MV/m. However, as the AC electric field rises to 1.6 MV/m, the condition $E < <E\pi/2$ is not maintained. This causes distortion in the AC component of the modulation signal PE, with asymmetric distortion introduced into the DC data calculated using the period average method. This impacts the AC field's influence on the DC output curve. To test this hypothesis, the maximum-derivative method is employed to extract the DC component from the sensor's output signal. By calculating the amplitude of a DC signal at the zero-crossing point of an AC signal, this method avoids the effects of asymmetrical distortion.

3.4 Typical application of sensors based on Kerr effect

3.4.1 In atmospheric air

The electric field sensing system illustrated in **Figure 10** consists of an 808 nm continuous wave (CW) laser, a series of optical components, including lenses, a polarizing beam splitter (PBS), quarter-wave plates, reflectors, and Bi4Ge3O12 (BGO) crystals, as well as analyzers and EOT-3010 (1.5 GHz) photodetectors (PD). Two BGO crystals are strategically positioned in the near-rod and near-plate regions within the electrode gap, allowing for simultaneous capture of the electric field in distinct corona regions. The employed BGO crystals have dimensions of $2 \times 2 \times 2$ mm and a 110-crystal orientation. Their compact size, relative to the gap distance, facilitates high-resolution electric field measurements with minimal interference from the crystals. This electro-optic (E-O) modulation technology is based on the Pockels effect. The laser beam, upon passing through the crystal, is modulated by the externally applied electric field, thereby encoding information proportional to the field strength. **Figure 11** presents the schematic of the E-O modulation process, wherein the modulated optical power is a function of the applied electric field:

$$P_o = \frac{P_{in}}{2}\left[1 + \sin\left(\frac{2\pi\gamma_{41}{n_0}^3 L E_z}{\lambda}\right)\right] \tag{10}$$

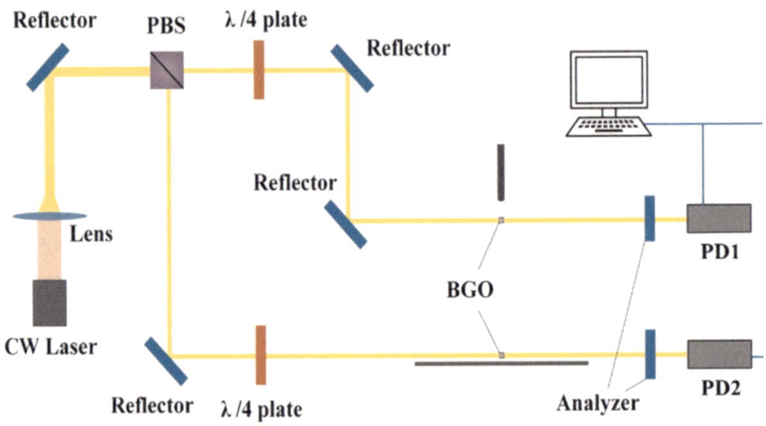

Figure 10.
Structure diagram of optical path for the electric field sensing system.

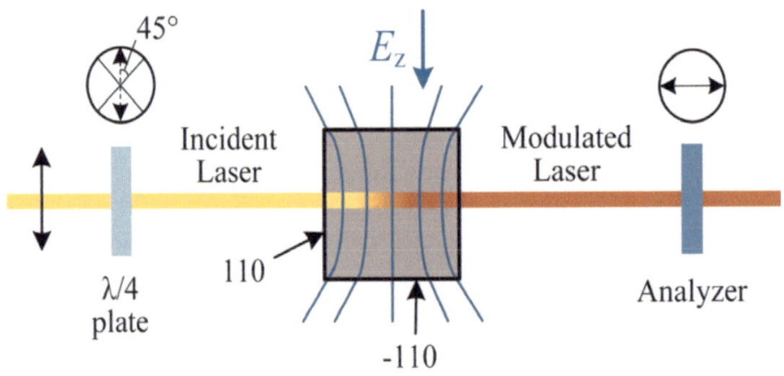

Figure 11.
Schematic diagram of electro-optic modulation based on BGO crystal.

In this context, Pin and Po represent the optical power of the input and output lasers, respectively, λ is the laser's wavelength, γ41 is the electro-optic coefficient of the BGO crystal, n0 is the crystal's natural refractive index, L denotes the optical path length through the crystal, and Ez is the projection of the external electric field in the -110direction.

Since the sinusoidal function in Eq. (2) remains small in practical applications, the optical power approximately maintains a linear relationship with the external electric field. Ultimately, the photodetector (PD) converts this optical power into a voltage signal, leading to the following relationship:

$$U_{\text{out}} = k_0 + k_1 E_z \tag{11}$$

Here, Uout is the output voltage from the PD, while k0 and k1 are determined by the optical system's parameters.

The system offers a rise time of 2.2 ns and operates linearly within a range of 0 to 1.4 MV/m, which is suitable for measuring corona discharge. The values of k0 and k1 can be experimentally determined, allowing Eq. (2) to calculate the electric field using Uout.

The experimental setup for this project is depicted in **Figure 12**. It primarily comprises a transformer with a maximum voltage of 80 kV, a protection resistor rated

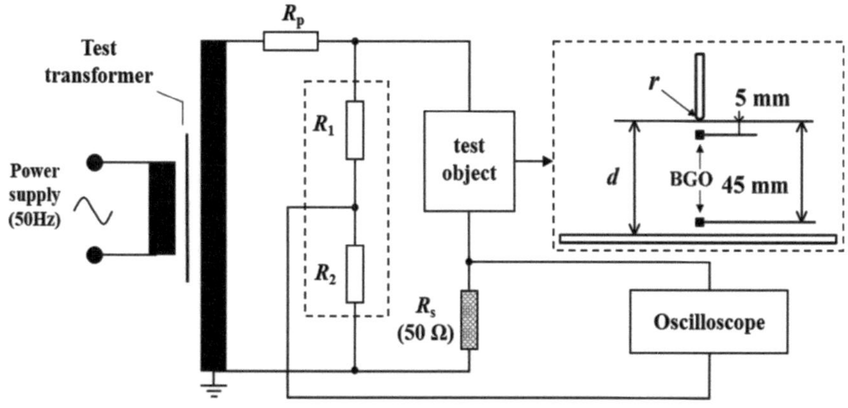

Figure 12.
Schematic diagram of the experimental setup.

at 1 MΩ, a rod-plate gap characterized by a rod electrode head radius (r) of 2.5 mm and a distance (d) of 50 mm between the rod and the plate, a voltage divider with a ratio of 1233:1, a sampling resistor valued at 50 Ω, and a Tektronix® MDO3012 oscilloscope with a bandwidth of 100 MHz. The voltage divider is arranged in parallel with the rod-plate gap to facilitate the measurement of the voltage across the gap, while the sampling resistor is connected in series to the gap to measure the current during corona discharge. Both the voltage divider and the sampling resistor are interfaced with the oscilloscope using 50 Ω cables. Tests have determined that the corona inception voltage for the rod-plate gap is approximately 8 kV; note that all voltage values are peak values of AC voltages.

Corona discharge under AC voltage demonstrates pronounced temporal periodicity and uneven spatial distribution, both of which are closely associated with the applied voltage. This section provides an analysis of the experimental results concerning the time-varying electric field for various discharge modes and spatial positions.

A power frequency voltage was applied across the rod-plate gap, with a BGO crystal positioned along the axis of the gap and 5 mm from the rod electrode. The electric field at this location was subsequently measured. **Figure 13** compares the sensing system's output in scenarios with and without corona discharge. The data reveals a correlation between the steps observed in the electric field and the current pulses generated by the corona discharge.

In **Figure 13(a)**, the applied voltage remains below the corona onset voltage for the gap, whereby only the Laplace field is prevalent in the space, resulting in the electric field being proportionally related to the applied voltage. The scale of the coordinate axis in the figure is adjusted to align the electric field with the applied voltage, thereby reflecting consistency with the voltage waveform.

Figure 13(b) elucidates the interrelation between the corona current and the spatial electric field. In this scenario, the applied voltage marginally exceeds the onset voltage, resulting in the observation of an electric field step. The electric field experiences rapid changes at specific instants, followed by relatively slow variations. The alterations in the electric field during the positive and negative half-cycles are asymmetrical, with the amplitude of the positive step being markedly larger than the negative step. Under this particular voltage condition, corona discharge presents instability, and not every cycle manifests an electric field step. Analyzing the current

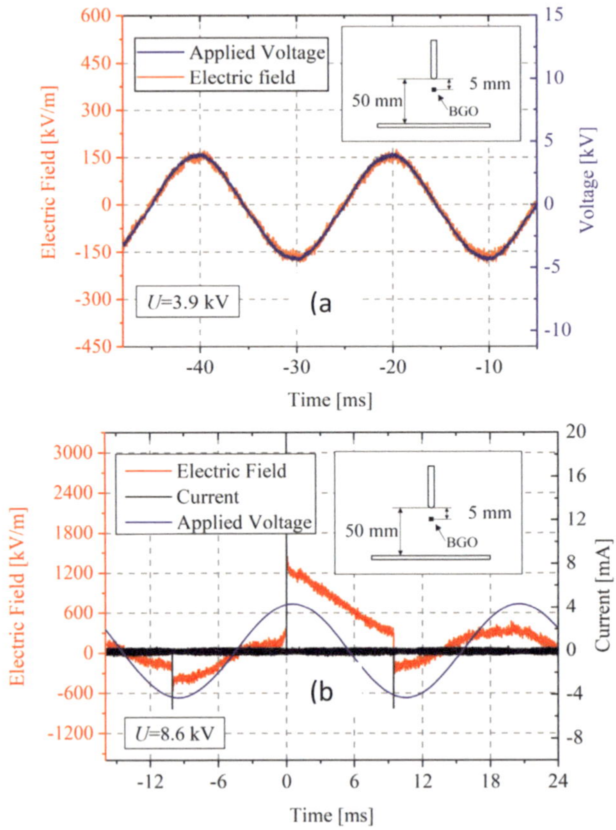

Figure 13.
Comparison between the Laplace field at low applied voltage and the corona discharge field at high voltage. (a) Laplace field at an applied voltage of 3.9 kV; (b) Corona discharge field at an applied voltage of 8.6 kV. Note: All voltages in this paper refer to the peak AC voltage.

signal alongside the electric field on the same temporal axis shows that each electric field step is accompanied by a corresponding current pulse. The current pulses observed during the positive half-cycle are robust, in contrast to the weaker pulses during the negative half-cycle, which concurs with established findings and previous research. Consequently, the steps in the electric field exhibit a correlation with the corona current pulses.

3.4.2 In medium oil

An experimental setup for generating AC/DC hybrid electric fields in oil was built to replicate the sensor's measurement environment in converter transformers, as illustrated in **Figure 14** [28].

Two parallel voltage generators are employed to deliver distinct types of voltage: a stable direct current (DC) voltage generator and an alternating current (AC) transformer operating at power frequency (**Figure 15**). Protection resistors are incorporated to prevent short circuits within the experimental chamber and to limit backflow between power sources. DC blocking capacitors effectively isolate the DC voltage, thereby reducing core bias in the transformer. By adjusting the outputs, three types of voltages can be applied: solely AC, solely DC, or a composite of AC and DC at varying

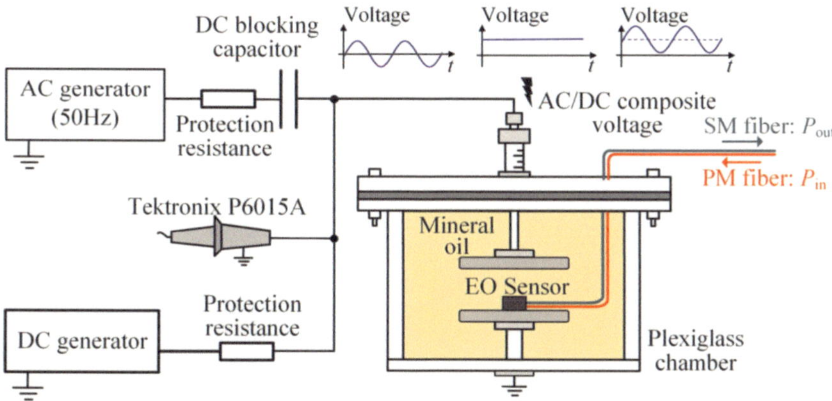

Figure 14.
Schematic diagram of the experimental setup.

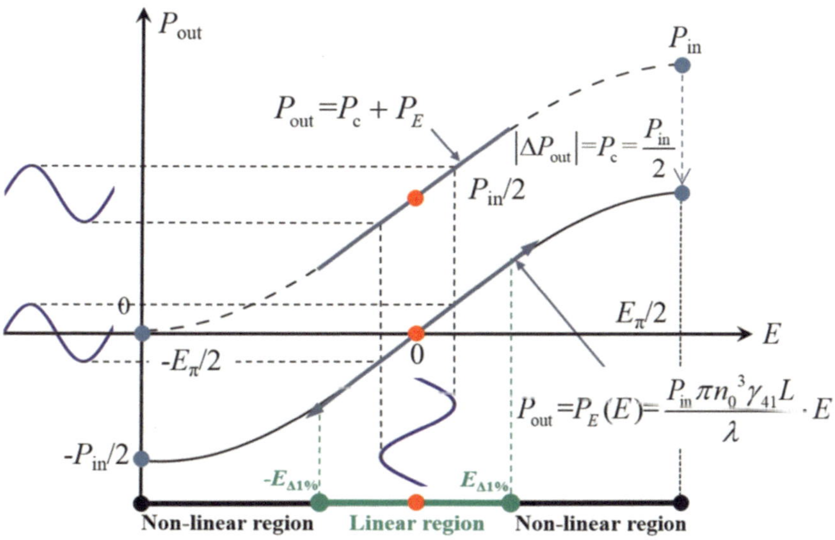

Figure 15.
Schematic of the sensing curve. In this context, the linear region is defined as the electric field range where the nonlinear error is less than 1%, [−EΔ1%, EΔ1%]. The nonlinear region is defined as the electric field region that causes a nonlinear error above 1%, [−Eπ/2, −EΔ1%] [EΔ1%, Eπ/2].

ratios. The applied voltage is measured using a Tektronix P6015A high-voltage probe, with specifications of 40 kV and 75 MHz. An integrated electro-optic (EO) sensor, equipped with two pigtails, is positioned within the gap to measure the electric field in oil. A linearly polarized laser is transmitted through the polarization-maintaining (PM) fiber to the EO sensor, and the resulting optical signal is conveyed to photodetectors (PD) *via* single-mode (SM) fiber. By studying the relationship between the applied electric field and the sensor's optical output, the output characteristics of the system can be analyzed (**Figure 16**).

The system's output waveforms under DC and AC fields are illustrated in **Figure 17**. As seen in **Figure 17(a)**, the DC bias is only 2.8 mV, representing a mere 0.2% of the original bias (Pc = 1.4 V), which demonstrates that the enhanced differential structure has substantially minimized Pc. When the applied field is 370 kV/m,

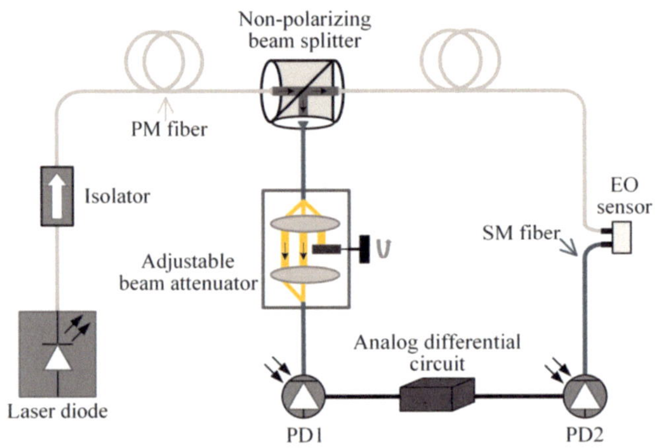

Figure 16.
Schematic diagram of dual-beam difference structure with adjustable optical power.

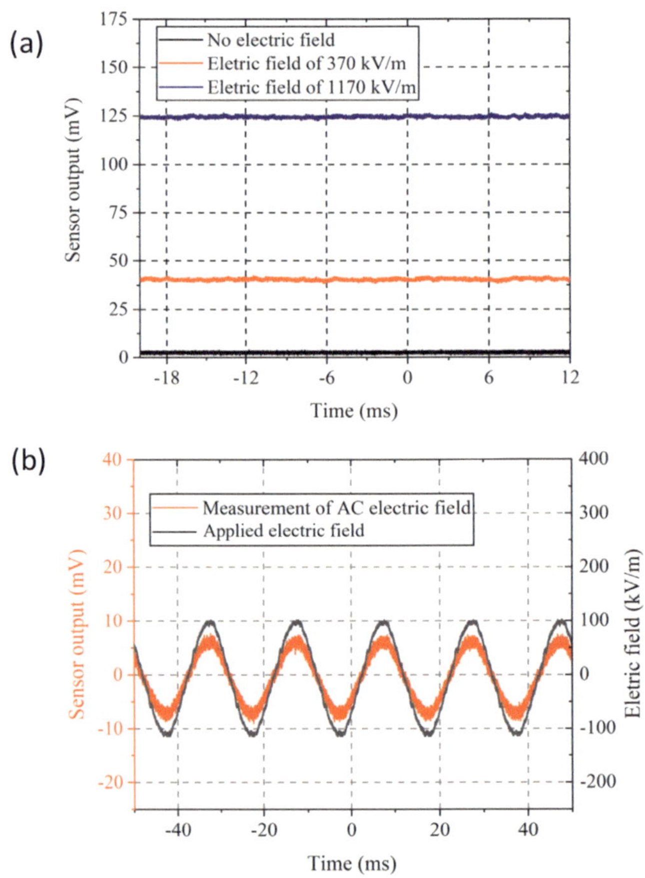

Figure 17.
EO sensor measurement waveform under AC and DC electric fields. (a) DC electric field measurement waveform and (b) AC electric field measurement waveform.

the output measures approximately 40 mV, whereas at an applied field of 1170 kV/m, the output measures about 124 mV. Consequently, the system's DC bias now exerts minimal influence on DC field measurements.

Figure 17(b) presents the AC electric field measurement results, where the waveform aligns with the applied field. Experiments indicate that the improved dual-beam differential method effectively suppresses Pc in the system output, yielding accurate DC field measurements. The system's capability to measure both AC and DC fields sets a foundation for examining EO sensor output characteristics under AC/DC hybrid fields.

Using the experimental setup, we obtained comprehensive output characteristics of the EO sensor in an oil medium under AC/DC hybrid fields, covering AC, DC, and various AC/DC hybrid ratios. The experimental findings here are supported by an analysis of the response mechanism.

4. Quantum sensors

4.1 Quantum electric field sensor

4.1.1 Electric field measurement based on NV color centers

When a diamond NV color center is exposed to a 532 nm laser and scanned with a specific microwave frequency, it emits a distinct fluorescence signal, which is used to characterize optical detection magnetic resonance (ODMR). In the absence of an external magnetic field, a single resonance peak at 2.87 GHz is observed in the spectrum. However, in the presence of an electric field, which induces a magnetic field, the degeneracy of the state ms = ± 1 is lifted, resulting in Zeeman splitting and producing two distinct resonance peaks at different frequencies [29]. The number and frequencies of these resonance peaks are critical for determining the direction of the magnetic field and the orientation of the NV axis. Additionally, the magnitude of the external electric field projected onto the NV axis can be derived from the resonance peak frequencies observed in the ODMR spectrum.

Compared with magnetic field measurement, NV color center is more difficult to measure electric field, mainly because the longitudinal magnetic field of NV color center will suppress its sensitivity to electric field. NV color center electric field sensor is mainly used for static field and low-frequency electric field measurement (kHz).

4.1.2 Electric field measurement based on Rydberg atoms

The principle of Rydberg atom electric field measurement relies on the interaction between Rydberg atoms and microwave fields [23]. This technique involves analyzing the splitting of Rydberg atomic energy levels to determine electric field strength. Rydberg atoms, particularly those in highly excited states with large principal quantum numbers, exhibit high sensitivity to external electric fields due to their long lifetimes, high polarizability, and large electric dipole moments, making them exceptional sensors for electric fields. When exposed to microwave fields, Rydberg atoms experience energy level splitting, which can be observed through quantum interference phenomena such as electromagnetically induced transparency and the Autler-Townes effect [11]. By measuring the extent of this splitting, the intensity of the microwave electric field can be accurately inferred, enabling precise electric field measurements.

In recent years, the EIT effect of Rydberg atoms has received widespread attention in the academic community, especially in the measurement of external fields (such as microwave fields, radio frequency fields, and terahertz fields).

4.2 Quantum magnetic field sensor

4.2.1 Optically pumped magnetometer technology

Optical pumping magnetometers (OPM) are typically classified into two categories based on their working elements: noble gas optical pumping magnetometers and alkali metal optical pumping magnetometers. Alkali metal optical pumping magnetometers utilize working substances such as potassium (K), rubidium (Rb), and cesium (Cs). In contrast, noble gas optical pumping magnetometers use noble gases like helium (He) and xenon (Xe). These devices are also referred to as atomic magnetometers due to their reliance on atomic properties [25].

The principle underlying the operation of OPM is the Zeeman effect, which describes the splitting of atomic energy levels in the presence of an external magnetic field. The magnetometers employ optical magnetic double resonance to detect the atomic Larmor frequency, which is influenced by the external magnetic field. Additionally, they utilize self-excited feedback oscillation to stabilize and lock the radio-frequency (RF) magnetic field frequency, enabling precise measurement of the magnetic field strength.

4.2.2 Coherent population trapping technique

Coherent Population Trapping (CPT) is a quantum interference effect realized in atomic systems [30]. The CPT atomic magnetometer is a quantum magnetometer based on the atomic coherent population trapping quantum effect and can achieve precise measurement of the magnetic field through CPT resonance. An obvious advantage of the CPT magnetometer over the OPM magnetometer is that it is based on all-optical resonance. Its magnetic probe does not rely on the radio frequency coil but only relies on optical components, making it possible to miniaturize the probe, thereby greatly improving the spatial resolution. At the same time, the additional magnetic field noise generated by the radio frequency coil can be completely avoided, making the CPT magnetometer have the characteristics of high accuracy, no measurement dead zone, and strong environmental adaptability.

4.2.3 Spin-free exchange relaxation technique

The spin-exchange relaxation-free (SERF) magnetometer is an advanced type of alkali metal atom magnetometer that operates in the SERF regime. Its principle involves utilizing a zero magnetic environment to significantly reduce the Larmor precession frequency of the atoms, while maintaining a high temperature to increase atomic density and, consequently, the spin-exchange rate. When the spin-exchange rate exceeds the Larmor precession frequency, the atomic spin-exchange relaxation is effectively suppressed, leading to enhanced precision in magnetic field measurements.

SERF magnetometers are the most sensitive magnetometers available, with the disadvantage that they can only operate in very weak magnetic fields and require expensive magnetic shielding.

4.2.4 Magnetic field measurement based on NV color centers

When the diamond NV color center is exposed to a 532 nm laser and scanned at a specific microwave frequency, it can generate a clear fluorescence signal, thereby obtaining the characteristics of optical detection magnetic resonance (ODMR). When there is no external magnetic field interference, only the 2.87 GHz resonance peak appears in the spectrum signal; under the influence of the magnetic field, the value of the degenerate state ms = ± 1 will change, resulting in Zeeman splitting, resulting in two resonance peaks with different frequencies. The number and frequency of the resonance peaks of the ODMR spectrum are important parameters for measuring the direction of the magnetic field opposite to the NV axis, and the magnitude of the external magnetic field projected onto the NV axis can be calculated by the resonance peak frequency of the ODMR spectrum [31].

Diamond NV color center systems are effective for magnetic measurements, with their capabilities depending on the concentration of NV centers. Specifically, NV color center ensembles are utilized for high-resolution magnetic measurements due to their collective signal averaging, which enhances spatial resolution. In contrast, single NV color centers are employed for high-sensitivity magnetic measurements, leveraging their quantum properties to detect minute variations in magnetic fields with exceptional precision.

At present, magnetometers based on diamond NV color centers can achieve sub-petersen-level magnetic field measurement sensitivity in low-frequency and high-frequency magnetic field environments.

5. Future scenario

This chapter provides a detailed overview of three types of sensors used for measuring electric and magnetic fields, reviewing their development, maturity, and specific characteristics.

Firstly, sensors based on the principle of electromagnetic induction, the earliest and most widely used, are noted for their simple structure and high sensitivity. However, they significantly impact the measured electromagnetic field. Secondly, optical effect-based sensors, which have rapidly evolved and become practical, feature a fully insulated structure, wide response frequency band, miniaturization, and strong anti-interference capability. Despite these advantages, their stability is heavily influenced by the performance of laser systems. Finally, quantum effect-based sensors, though still in the early stages of development, exhibit remarkable performance benefits. Their sensitivity is more than three orders of magnitude greater than that of the previous two types, making them highly promising for future advancements.

As discussed in the introduction, all measurement methods inevitably disturb the parameters being measured. Therefore, when selecting electromagnetic field sensors for practical applications, it is crucial to understand the interaction between the sensor and the measured parameters to identify the optimal choice. For instance, electromagnetic induction sensors are best suited for open-space measurements such as electromagnetic environment detection of transmission lines, lightning overvoltage measurements, and communication station radiation assessments. Optical effect sensors, with their compact size and insulation, are ideal for harsh electromagnetic environments and confined spaces. Quantum effect sensors, with their exceptional sensitivity, are expected to excel in applications like electric field calibration and monitoring extremely weak electric fields.

In conclusion, this chapter has covered widely used electromagnetic induction-based sensors, rapidly advancing optical effect-based sensors, and emerging quantum effect-based sensors. The aim is to provide a thorough understanding of electromagnetic measurement techniques, facilitate their application in various scientific and engineering fields, and contribute to technological progress.

Acknowledgements

The author acknowledges the usage of Chat Generative Pre-trained Transformer for language polishing of the manuscript.

Author details

Xing Fan[1], Zhehao Pei[1]* and Qingrong Peng[2]

1 Hefei University of Technology, Anhui, China

2 North China Electric Power University, Beijing, China

*Address all correspondence to: peizhehao@126.com

References

[1] Long Z, Zhou F, Lin F, Fan J, Li W, Diao Y, et al. Development of a wideband precision electric field measuring sensor. Sensors (Basel). 25 Nov 2023;**23**(23):9409. DOI: 10.3390/s23239409. PMID: 38067780; PMCID: PMC10708871

[2] Araujo JFDF, Junior EBM, Mendoza LAF. A simple portable magnetometer based on magnets and hall-effect sensors capable of measuring magnetic properties. Applied Sciences. 2022;**12**(24):12565-12565

[3] Vadde A, Meghana TV, Sahana KM, Keerthi TG, Gouthami Y, et al. Investigation of magnetic flux density on conducting and non-conducting cylindrical surface using magnetic sensor. Materials Today: Proceedings. 2022;**65**(P8):3623-3628

[4] Blagojević M, Jovanović U, Jovanović I, et al. Realization and optimization of bus bar current transducers based on hall effect sensors. Measurement Science and Technology. 2016;**27**(6):065102

[5] Chungang Z, Wei L, Hao Z, et al. Power frequency electric field measurement based on electromagnetic induced transparent spectrum under radio frequency field. Acta Photonica Sinica. 2021;**50**(6):154

[6] Zhiqiang C, Weibin C, Hailun Z, et al. A model to design giant magnetoresistive sensor. Journal of Physics: Conference Series. 2021;**1739**(1):012037

[7] Jia J, Lin J. High-voltage broadband high-current sensor and its anti-interference design based on Giant Magnetoresistive effect. In: Proceedings of 2018 International Conference on Computational Science and Engineering (ICCSE 2018). Jiangsu Electric Power Company Taizhou Power Supply Company, International Information and Engineering Association; Jiangsu Xinyidi Smart Technology Co., Ltd; 2018. p. 3

[8] Weiss XF et al. Simple mathematical operation-based calibration method for giant magnetoresistive current sensor applying B-spline modeling. IEEE Sensors Journal. 2016;**16**(12):4733-4739

[9] Jianlei H, Linghao C, Zhenzhen G, et al. Sensitivity enhanced magnetic field sensor based on Farady effect and dual-polarization fiber grating laser. Optics Express. 2013;**21**:30156-30162

[10] Gao C, Zhao X, Zhang S, et al. Influencing factors and uncertainty analysis for Kerr electro-optic effect based electric field measurements in transformer oil under impulse voltage. Measurement Science and Technology. 2024;**35**(4):045202

[11] Ling L, Yucchun J, Jinlian H, et al. Super low-frequency electric field measurement based on Rydberg atoms. Optics Express. 2023;**31**(18):29228-29234

[12] Jia-Wei Z, Xuan M, Xiaofei W, et al. An optical sensor with wide measurement range for the magnetic field detection. Sensors and Actuators: A. Physical. 2023;**363**:114757

[13] Deng M, Liu D. Multilongitudinal mode fiber laser for highly-sensitive magnetic field measurement. Journal of Optics. 2016;**18**(10):105802-105802

[14] Dai J, Yang M, Li X, et al. Magnetic field sensor based on magnetic fluid clad etched fiber Bragg grating. Optical Fiber Technology. 2011;**17**(3):210-213

[15] Ryouto I, Kazuhiko O. Development of static load sensor utilizing both direct and inverse piezoelectric effect of piezoelectric element. The Proceedings of Mechanical Engineering Congress, Japan. 2018;**2018**:G1500203-G1500203

[16] Wu B, Chen H, Wang M, et al. High resolution and large measurement range voltage sensing based on an optoelectronic oscillator utilizing an unbalanced Mach-Zehnder interferometer. Optics Express. 2024;**32**(5):8030-8041

[17] Allen RG. Kerr Voltage Sensor. United States: Sandia National Laboratories; 1986

[18] Xiao Y, Fan X, Pei Z, Chen W, Zhang Q. Mechanism analysis of thermal-induced periodic fluctuations in electro-optic sensor. IEEE Transactions on Instrumentation and Measurement. 2024;**73**:1-7. Art no. 9500507. DOI: 10.1109/TIM.2023.3331409

[19] Fan X, Chen W, Pei Z, Wen T, Wu Z, Zhang Q. Evolution characteristics of electric field for AC Corona modes in atmospheric air. IEEE Transactions on Plasma Science. 2021;**49**(1):219-225. DOI: 10.1109/TPS.2020.3043597

[20] Xing F, Weijiang C, Qiaogen Z, et al. An electric field measurement method based on electro-optical modulation for corona discharge in air. The Review of Scientific Instruments. 2019;**90**(8):084704

[21] Zong Da Z, SiYu Y, LiCheng W, et al. Single NV centers array preparation and static magnetic field detection. Optics Express. 2022;**30**(18):32355-32365

[22] Horsthemke L, Pogorzelski J, Stiegekötter D, et al. Excited-state lifetime of NV centers for all-optical magnetic field sensing. Sensors. 2024;**24**(7):2093

[23] Xiao D, Shi Z, Chen L, et al. Low-frequency weak electric field measurement based on Rydberg atoms using cavity-enhanced three photon system. Frontiers in Physics. 2024;**12**:1405149

[24] Gavrilenko VP, Kim HJ, Ikutake T, Kim JB, Choi YW, Bowden MD, et al. Measurement method for electric fields based on stark spectroscopy of argon atoms. Physical Review. E. 2000;**62**(5 Pt B):7201-7208

[25] Ziqi Y, Ying L, Min X, et al. Compact multi-channel optically pumped magnetometer for bio-magnetic field imaging. Optics and Laser Technology. 2023;**164**:109534

[26] Rushton LM, Pyragius T, Meraki A, Elson L, Jensen K, et al. Unshielded portable optically pumped magnetometer for the remote detection of conductive objects using eddy current measurements. The Review of Scientific Instruments. 2022;**93**(12):125103-125103

[27] Troullinou C, Jiménez-Martínez R, Kong J, Lucivero VG, Mitchell MW, et al. Squeezed-light enhancement and backaction evasion in a high sensitivity optically pumped magnetometer. Physical Review Letters. 2021;**127**(19):193601-193601

[28] Fan X, Chen W, Guo C, Wu Z, Pei Z, Zhang Q. Sensitivity characteristics of electro-optical sensors under AC/DC hybrid electric fields in oil medium environment. IEEE Transactions on Instrumentation and Measurement. 2021;**70**:1-12. Art no. 1504312. DOI: 10.1109/TIM.2021.3096272

[29] Hollendonner M et al. Quantum sensing of electric field distributions of

liquid electrolytes with NV-centers in nanodiamonds. New Journal of Physics. 2023;**25**(9):093008

[30] Ermak VS, Podvyaznyi AA, Semenov VV. Multiple frequency radiofrequency-optical resonance in a four-level model of rubidium atoms with optical pumping. Journal of Applied Spectroscopy. 2006;**73**(3):366-370

[31] Timo W, Christian G, Florian F, et al. Determination of the three-dimensional magnetic field vector orientation with nitrogen vacany centers in diamond. Nano Letters. 2020;**20**(5): 2980-2985

Chapter 5

Electrostatics in Engineering Photodiodes for Earth Observation Cameras

Sri Harsha Kodati and Manisha Muduli

Abstract

Electrostatics plays a pivotal role in designing semiconductor devices such as transistors, capacitors, solar cells, power devices, and photodetectors. This chapter delves into the critical applications of electrostatics in designing photodetectors for cameras used in earth observation satellites. Earth observation satellites are crucial for environmental monitoring, resource management, disaster response, and scientific research. This chapter focuses on three case studies that illustrate the use of electrostatic principles to design photodetectors for specific applications: (1) Visible cameras for high-resolution imaging (2) Short-wave infrared cameras for precision farming (3) Mid-wave infrared cameras for environmental monitoring. Each case study presents unique technical requirements that can be met through meticulous semiconductor device engineering, leveraging the principles of electrostatics. Using a first principles approach, the chapter begins with the charge, electric field, and potential distributions in a simple P-N junction diode, which forms the basis for most photodetectors. The following sections extend to how these distributions can be manipulated to design photodetectors tailored to each application. By the end of this chapter, readers will understand how to use the fundamentals of electrostatics to design advanced photodetectors based on specific technical requirements for earth observation.

Keywords: P-N junction photodetectors, earth observation (EO) satellites, visible (VIS) cameras, short-wave infrared (SWIR) cameras, mid-wave infrared (MWIR) cameras, electrostatic engineering

1. Introduction

Electrostatics forms the foundation of semiconductor physics, playing a critical role in the design of photodetectors for earth observation (EO) satellites. These satellites are indispensable for applications such as environmental monitoring, disaster response, and resource management, utilizing specialized cameras to capture data across distinct spectral bands. Visible light cameras are used for general high-resolution imaging, short-wave infrared (SWIR) cameras for precision farming, and mid-wave infrared (MWIR) cameras for environmental monitoring [1, 2]. The performance of these photodetectors depends on meticulous electrostatic and material design choices that directly influence key metrics at both the sensor and satellite levels.

Figure 1 presented here illustrates the multi-level design considerations, from satellite-level requirements down to the photodetector-level parameters that can be manipulated through electrostatic and material engineering. At the satellite level, key the most important metrics are *speed*, *performance*, and *cost*. These high-level metrics flow down to the camera level, where *frame rate (or line rate for line scanner cameras)*, *signal-to-noise ratio* (SNR), and *system size* (mass and volume) become critical factors. Signal quality (or SNR) is influenced by the quantum efficiency (QE), while noise is primarily determined by the dark current. Additionally, operating temperature affects both mass and cost, as lower temperatures require cooling systems that add to the satellite's size and complexity. Pixel size is a crucial factor; smaller pixels reduce telescope size and weight for a given resolution which reduces the cost, but smaller pixels increase surface dark current, which in turn adds to the noise [3].

On the photodetector level, the important figure of merits are *bandwidth*, *QE*, and various *dark current components* (diffusion, Shockley-Read-Hall (SRH) recombination, trap-assisted tunneling (TAT), band-to-band (BTB) tunneling, and surface dark current) [4]. These parameters collectively shape the camera's frame rate, SNR, and overall performance, and can be fine-tuned through electrostatic and material design adjustments. Note that this flow-down emphasizes the influence of electrostatic and material design parameters, which is why it does not address other factors that impact satellite speed, performance, and cost.

In the subsequent sections, the chapter explores the basics of P-N junction photodiodes, and the photodetector-level parameters are explored in detail, examining the governing equations, and identifying the key variables that influence them. Some of these variables—such as doping concentration, thickness, device area, time delay integration, barriers, and field engineering—can be adjusted independently of the chosen active sensing material, allowing the engineers to fine-tune the electrostatics of

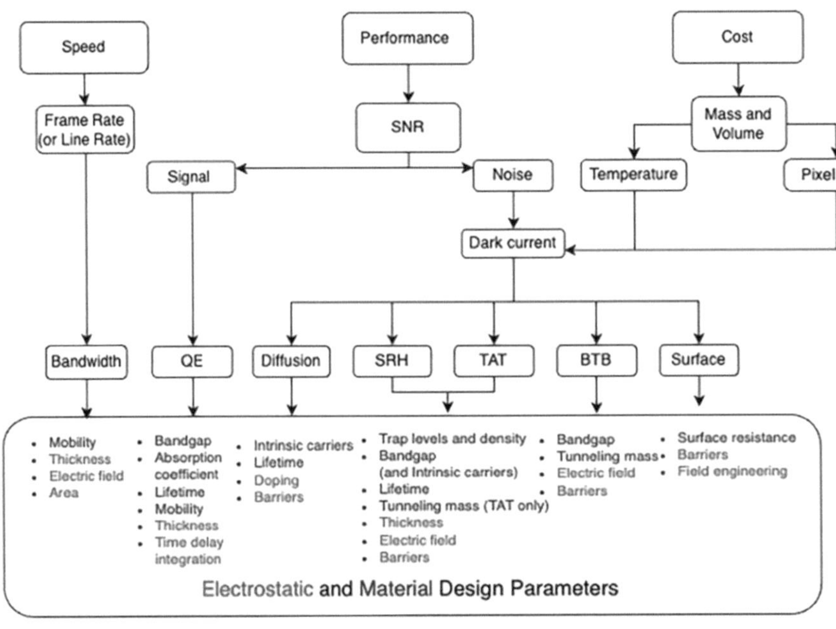

Figure 1.
Impact of electrostatic (green) and material (blue) dependent parameters on satellite imaging performance.

the device. Other variables are specific to the sensing material itself. Further, the chapter discusses how manipulating these critical variables enhances photodetector-level attributes, thereby improving camera and satellite capabilities in terms of speed, efficiency, and cost-effectiveness.

The final section concludes with case studies, illustrating how electrostatics can be applied to optimize photodetectors for different EO applications in VIS, SWIR, and MWIR spectral ranges.

2. P-N junction photodiodes

Most recent photodiodes are P-N junctions with three main regions: a neutral P region, a neutral N region, and a depletion region [5–11]. The depletion region forms as electrons and holes diffuse across the junction, leaving behind immobile charges (negative charges in the P region and positive charges in the N region) as shown in **Figures 2(a)** and **2(b)**. These charges then affect the electric field, potential, and energy band diagram of the photodiode.

i. Charge to the electric field: These immobile charges create an internal electric field. By Gauss's law, the electric field E in the depletion region relates to charge density ρ by

$$\frac{dE}{dx} = \frac{\rho}{\in} \tag{1}$$

Here, ϵ is the permittivity of the semiconductor material. Integrating this equation across the depletion region gives the built-in electric field profile within the junction, as shown in **Figure 2(c)**, whose magnitude reaches a maximum at the junction and decreases toward the edges.

ii. Electric field to potential: The electric field generates a potential difference V_j (the built-in voltage) across the depletion region, given by,

$$E = -\frac{dV}{dx} \tag{2}$$

Integrating this equation across the depletion region gives the built-in potential distribution as shown in **Figure 2(d)**.

iii. Potential to energy band diagram: The built-in electrostatic potential $V(x)$ across the junction affects the energy levels of electrons and holes, as shown in **Figure 2(e)**. The relationship between the electrostatic potential and the energy band edges can be expressed as:

$$E_c(x) = E_{co} - qV(x) \tag{3}$$

$$E_v(x) = E_{va} - qV(x) \tag{4}$$

Where, $E_c(x)$ and $E_v(x)$ represent the conduction band and valence band energy levels across the junction, respectively. E_{co} and E_{va} are the conduction and valence band edges far from the junction. q is the charge of an electron. The intrinsic level E_i

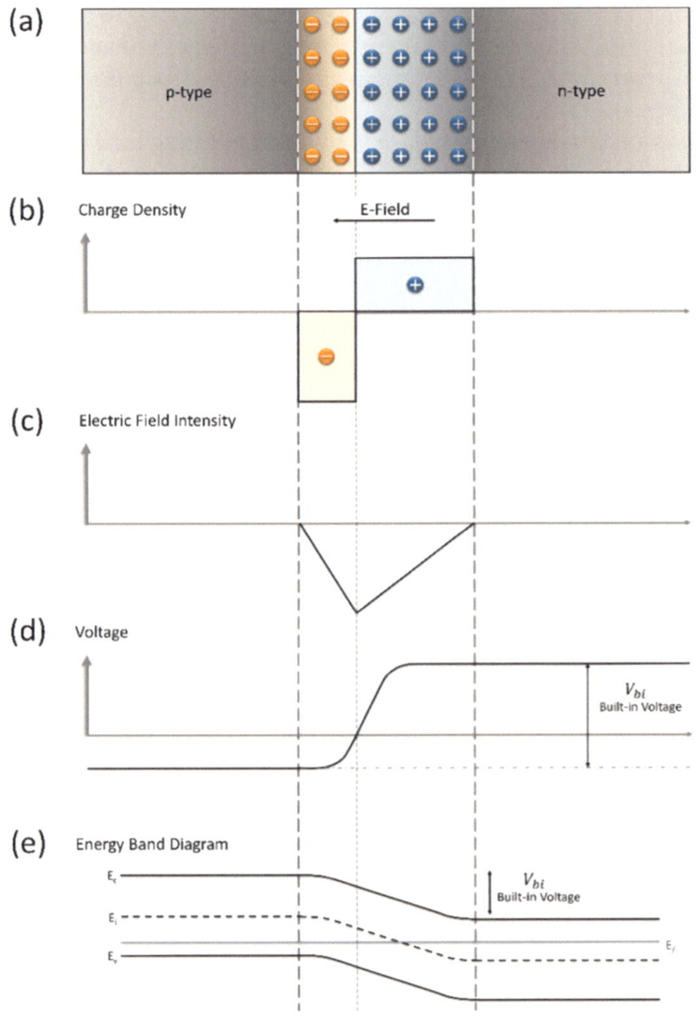

Figure 2.
(a) Illustration of a simple P-N junction at equilibrium. (b) Charge density, (c) internal electric field, (d) built-in voltage, and (e) energy band diagram of the P-N junction at equilibrium.

lies around midway between the conduction and valence bands in an intrinsic semiconductor, while the Fermi level E_f represents the energy with a 50% probability of electron occupancy.

When a reverse-bias voltage (V_a) is applied to a P-N junction photodiode, the P-region is connected to the negative terminal, and the N-region to the positive terminal of an external voltage source. As an external voltage is applied, the charge, field, and potential diagrams change, resulting in the non-equilibrium energy band diagram shown in **Figure 3**.

Due to the applied reverse bias V_a, the fermi level on the P and N sides split until they reach a steady state where the difference in these energy levels is V_a. This reverse bias increases the total voltage drop, which becomes the sum of the built-in potential V_{bi}, and the applied reverse bias V_a. The increased potential difference across the junction enhances the electric field in the depletion region. A stronger electric field

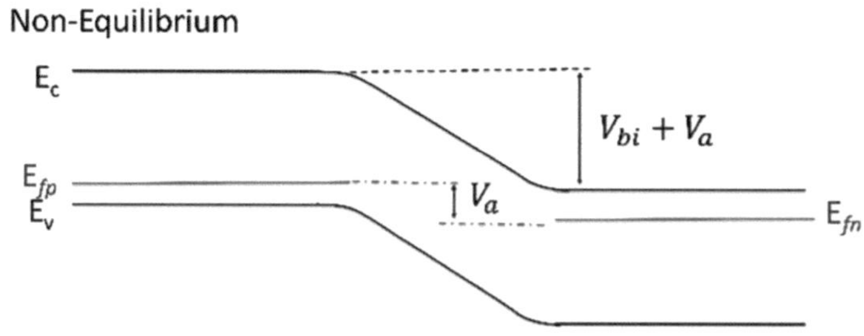

Figure 3.
Non-equilibrium reverse bias band diagram of a P-N junction.

pushes the majority carriers away from the junction, widening the depletion region and suppressing the majority carrier flow across the junction. As a result, the reverse current is dominated by the movement of minority carriers and other thermally generated carriers, collectively known as the *dark current*. This dark current contributes to the noise portion of the signal-to-noise ratio (SNR) of a photodiode.

Under reverse bias, when a photon of sufficient energy strikes the photodiode, it generates electron-hole pairs through the photoelectric effect. The strong electric field in the depletion region quickly separates these photogenerated carriers, sweeping electrons toward the N-region (anode) and holes toward the P-region (cathode). This separation of carriers produces a *photocurrent* whose magnitude is directly related to the *Quantum Efficiency (QE)*—the ability of the photodiode to convert incoming light into photocurrent. Maximizing QE enhances photocurrent and, therefore, signal strength. The QE (or the photocurrent) contributes to the signal part of the SNR of a photodiode, which is given by,

$$SNR = \frac{I_p^2}{\langle i_{shot}^2 \rangle + \langle i_j^2 \rangle} \tag{5}$$

Where, I_p is the photocurrent, which is a function of QE, i_{shot} is the shot noise, generated due to both dark current and photocurrent in the detector, and i_j is the Johnson-Nyquist noise or resistor noise that is inversely proportional to the photodiode resistance. The crucial parameters for a high SNR are a higher QE and a lower dark current.

Another critical photodiode parameter is *bandwidth*, which represents the speed at which the photodiode can convert optical input into electrical output. Higher bandwidth allows the photodiode to respond to rapid changes in light intensity, making it suitable for high-speed applications.

3. Material-dependent and electrostatic design parameters of a photodiode

The effectiveness of photodiodes can be evaluated through these three key parameters—QE, dark current, and bandwidth. Together, these parameters influence the camera's SNR, size, and frame rate (or line rate), which in turn impact the performance, cost, and speed of satellite systems tailored to specific applications.

3.1 Quantum efficiency

Quantum Efficiency (QE) is the measure of the signal part of the photodiode, the photocurrent. QE is the ratio of the number of electron-hole pairs collected to the number of incident photons, quantifying how effectively the photodiode converts incoming light into an electrical signal. High QE is desired for maximizing signal strength and sensitivity.

$$QE = (1 - R)(1 - e^{-\alpha t})\eta_c \tag{6}$$

The QE equation has three main parts:

i. Radiation incident on the detector $(1 - R)$: This term accounts for the fraction of light that enters the photodiode after reflection losses, quantified by the reflection coefficient R. Minimizing R increases the effective amount of light available for absorption and conversion to photocurrent.

ii. Portion of light absorbed $(1 - e^{-\alpha t})$: This term describes the fraction of light absorbed within the active region. It is quantified by t, the thickness of the active region of the photodiode, and α, the absorption coefficient. The absorption coefficient depends on the material's bandgap and the wavelength of incoming light. For a given wavelength range (e.g., VIS, SWIR, MWIR), the photodetector material must have an appropriate bandgap to maximize absorption. The relationship between the *bandgap energy* E_g and the application's *target wavelength of* λ can be approximated by:

$$E_g = \frac{hc}{\lambda} \tag{7}$$

Where h = planks constant, c = speed of light, λ = target wavelength. This relation shows that materials with different bandgaps are chosen based on the target wavelength range: VIS, SWIR, or MWIR.

iii. Carrier generation and transport (η_c): This part represents the efficiency with which absorbed photons generate carriers that reach the electrodes. The carrier collection efficiency η_c depends on the *lifetime* of the carriers (how long they exist before recombination) and their *mobility* (how quickly they can move under the influence of an electric field). Higher mobility and longer carrier lifetimes increase η_c improving the QE by ensuring that more generated carriers contribute to the photocurrent.

Crucial variables to maximize QE: *Higher Absorption Coefficient* α for the target wavelength range, achieved by selecting a material with an appropriate bandgap. Increased thickness t of the active region, allowing more radiation to be absorbed. *Higher carrier mobility and lifetime* to enhance η_c enabling efficient collection of generated carriers. While t helps fine-tune the electrostatics of the device independent of the active sensing material, the other variables are sensing material dependent.

3.2 Bandwidth

Bandwidth refers to the speed at which a photodiode can respond to changes in light intensity, which relates to the framerate of the camera. Higher bandwidth allows the photodiode to capture rapid variations in light, making it suitable for high-speed applications. Bandwidth is quantified as the 3 dB bandwidth, representing the frequency range over which the detector can effectively operate. It is defined as the maximum rate at which a detector can generate output signal with respect to its input optical signal and is measured from DC to cut-off frequency. It is denoted as f_{3dB},

$$f_{3dB} \approx \sqrt{\frac{1}{\frac{1}{f_{RC}^2} + \frac{1}{f_t^2}}} \approx \sqrt{\frac{1}{\frac{1}{\left(\frac{W}{2\Pi R_s \in A}\right)^2} + \frac{1}{\left(\frac{\mu E}{2\Pi W}\right)^2}}} \tag{8}$$

The bandwidth equation has two main components:

i. RC-limited bandwidth: Represented by the term f_{RC}, this component depends on the junction capacitance (C) and series resistance of the photodiode (R_s), that originates from semiconductor–metal contact interface. The $\mathbf{C}$ is influenced by the dielectric constant $\in$, *area* A and *depletion width* W.

ii. Transit time-limited bandwidth: Represented by the term f_t, this component quantifies the speed at which carriers traverse the depletion region, quantified by mobility μ, the electric field E across the depletion region, and the depletion region thickness W.

Crucial variables to maximize bandwidth: *Lower* R_s and a *smaller* A reduces the RC time constant, improving f_{RC}. *Higher* μ and *stronger* E increase the speed of carrier collection, improving f_t. While a *larger* W benefits the RC-limited bandwidth, a smaller W is preferable for the transit time-limited bandwidth, creating a trade-off. While W, A, *and* E help fine-tune the electrostatics of the device independent of the active sensing material, μ is a sensing material-dependent parameter.

3.3 Dark current

Dark current is the leakage current that flows through a photodiode even when no light is present, caused by the diffusion of minority carriers within the quasi-neutral regions, thermally generated electron–hole pairs, and tunneling. This current contributes to noise, reducing the photodiode's signal-to-noise ratio (SNR).

The equivalent circuit model of a P-N junction photodiode is shown in **Figure 4**, illustrating the various components that contribute to dark current.

$$J_{Total} = J_D + J_{GR} + J_{TAT} + J_{BTB} + J_{Sh} \tag{9}$$

Where, the total current density, J_{Total} (dark current normalized with photodiode area), is originating from diffusion (J_D), generation-recombination (J_{GR}), trap assisted tunneling (J_{TAT}), band-to-band tunneling (J_{BTB}), and from surface components of the photodiode (J_{Sh}). Additionally, the relationship between the applied voltage (V_a), the junction voltage (V_j), and the series resistance (R_s) is given by:

$$V_a = V_j + R_s I \tag{10}$$

Where I is the total dark current. For an ideal semiconductor–metal interface, the R_s can be ignored. R_{sh} is the shunt resistance at the surface of the photodiode.

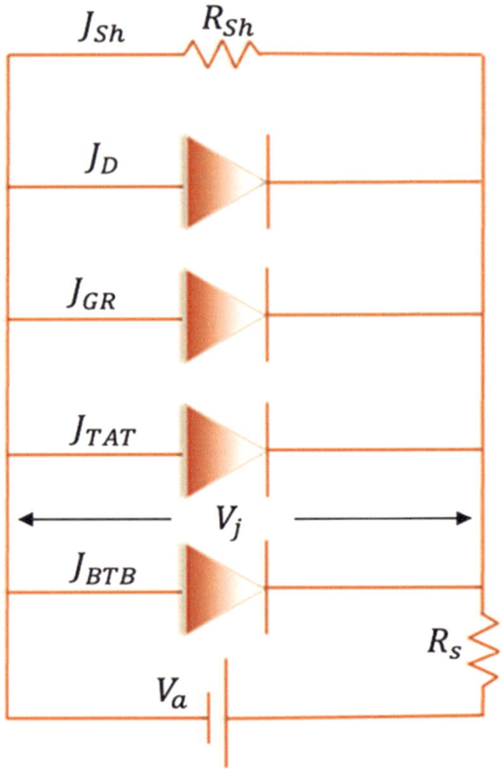

Figure 4.
Dark current components in the equivalent circuit model of a P-N junction photodiode.

i. Diffusion dark current (J_D): The diffusion dark current originates from the minority carriers diffusing in the quasi-neutral part of the P and N regions. When a reverse bias is applied to the photodiode, minority electrons in the P region move toward the depletion region through diffusion. Once in the depletion region, these electrons are transported by the internal electric field. Similarly, minority holes in the N region diffuse toward the depletion region, where they are also carried by the electric field. The J_D equation is,

$$J_D = q\left(\frac{D_n n_i^2}{L_n N_A} + \frac{D_p n_i^2}{L_p N_D}\right)\left(\exp\left(\frac{qV_j}{k_B T}\right) - 1\right) \tag{11}$$

Here, D_n and D_p are the diffusion coefficients for electrons and holes, respectively, L_n and L_p are the diffusion lengths for electrons and holes, respectively, N_A and N_D are the acceptor and donor doping concentrations, respectively, k_B is Boltzmann's constant, and T is the operating temperature in Kelvin. n_i is the intrinsic carrier concentration given by,

$$n_i = N_c N_v \exp\left(-\frac{E_g}{2k_B T}\right) \tag{12}$$

Where N_c and N_v is the effective density of states in the conduction and valance bands, respectively.

Crucial variables: It can be understood that the most influenceable variables to reduce J_D are a lower n_i (higher E_g and a lower T), and a higher N_A and N_D. While n_i is sensing material dependent, N_A and N_D help fine-tune the electrostatics of the device independent of the active sensing material. In addition, barrier structures can also be used to manipulate the electrostatics to block the diffusion current that will be discussed in Section 4.1.1.

ii. GR dark current (J_{GR}): The GR dark current originates from the thermal generation of carriers within the depletion region. Under reverse bias, the depletion region widens, creating conditions where the G-R current is more prominent. First, compared to the quasi-neutral regions of the P-N junction, the depletion region has a much lower density of free carriers. In this low-carrier-density environment, traps—energy levels within the bandgap introduced by defects or impurities, can capture and release carriers more effectively. These trap states, typically located near the mid-gap energy, act as intermediate sites for electrons and holes, facilitating the trap-assisted carrier generation process.

Second, in the depletion region, the Fermi energy level (the energy with a 50% probability of electron occupancy.), is typically positioned closer to the mid-bandgap trap level (or the intrinsic energy level), resulting in a roughly 50% probability that traps will be occupied in a steady state. This balanced occupancy allows traps to capture and release carriers with equal likelihood, enabling continuous electron–hole transitions. In contrast, in the quasi-neutral regions, the Fermi level is closer to the conduction band (n-side) or valence band (p-side), making trap levels either mostly filled or mostly empty. Together, these factors—low carrier density and balanced trap occupancy—make trap-assisted generation, the dominant source of G-R dark current in reverse bias. The J_{GR} equation is,

$$J_{GR} = \frac{2n_i W k_B T}{(V_{bi} - V_j)\tau_{GR}} \cdot \sinh\left(\frac{qV_j}{2k_B T}\right) \cdot f(b) E_g = \frac{hc}{\lambda} \quad (13)$$

$$f(b) = \frac{1}{2\sqrt{b^2 - 1}} \ln\left(2b^2 + 2b\sqrt{b^2 - 1} - 1\right) E_g = \frac{hc}{\lambda} \quad (14)$$

for b > 1

$$f(b) = 1 \quad (15)$$

for b = 1

$$f(b) = \frac{1}{\sqrt{1 - b^2}} \tan^{-1}\left(\frac{\sqrt{1 - b^2}}{b}\right) \quad (16)$$

for b < 1

$$b = e^{-qV/2k_B T} \cosh\left(\frac{E_t - E_i}{k_B T}\right) \quad (17)$$

Here, W is the width of the depletion region, V_{bi} is the built-in potential, τ_{GR} is the generation-recombination lifetime (that is function of trap densities), E_t is the trap energy level, and E_i is the intrinsic energy level.

Crucial variables: To reduce J_{GR}, key variables include a lower W, a higher τ_{GR}, a lower n_i (higher E_g and a lower T). While n_i and τ_{GR} are sensing material dependent, W helps fine-tune the electrostatics of the device independent of the active sensing material. In addition, barrier structures can also be used to manipulate the electrostatics to block the GR current that will be discussed in Section 4.1.2.

iii. TAT dark current (J_{TAT}): Trap-assisted tunneling occurs when carriers use intermediate trap energy levels within the bandgap to quantum mechanically tunnel through the depletion region, particularly under a high reverse-bias. These traps act as steppingstones, allowing carriers to move from the valence band to the conduction band (or vice versa) in multiple small steps instead of overcoming the full energy barrier (the bandgap) in one go. The J_{TAT} equation is,

$$J_{TAT} = \frac{q^2 m_t M^2 N_t V_j}{8\pi\hbar^3 (E_g - E_t)} \exp\left(-\frac{4\sqrt{2m_t (E_g - E_t)^3}}{3qE\hbar}\right) \tag{18}$$

$$E = \sqrt{\frac{2q(V_{bi} - V_j) N_A N_D}{\epsilon(N_A + N_D)}} \tag{19}$$

$$V_{bi} = \frac{kT}{q} \ln\left(\frac{N_A N_D}{n_i^2}\right) \tag{20}$$

Here, m_t is the effective mass for tunneling, M^2 is the transition matrix element for tunneling, N_t is the trap density, and $\hbar$ is the reduced Planck's constant.

Crucial variables: To reduce J_{TAT}, key variables include reducing the E, N_t, and m_t, and increasing the E_g. While N_t, m_t, and $E_g$are sensing material dependent, E helps fine-tune the electrostatics of the device independent of the active sensing material. E can also be reduced by increasing the thickness of the depletion region for a given applied voltage. In addition, barrier structures can also be used to manipulate the electrostatics to block the J_{TAT}, that will be discussed in Section 4.1.3.

iv. BTB dark current (J_{BTB}): Band-to-band tunneling occurs when electrons can quantum-mechanically tunnel directly from the valence band to the conduction band across a narrow depletion region under high electric fields. This process is independent of trap states and relies purely on the strong electric field to overcome the bandgap. The J_{BTB} (is,)

$$J_{BTB} = \frac{q^3 E V_j \sqrt{2m_e}}{4\pi^2 \hbar^2 \sqrt{E_g}} \exp\left(-\frac{4\sqrt{2m_t E_g^3}}{3qE\hbar}\right) E_g = \frac{hc}{\lambda} \tag{21}$$

Crucial variables: To reduce J_{BTB}, key variables include reducing the E, and m_t, and increasing the E_g. While m_t and $E_g$are sensing material dependent, E helps fine-tune the electrostatics of the device independent of the active sensing material. E can be engineered by changing the thickness of the depletion region. In addition, barrier structures can also be used to manipulate the electrostatics to block the J_{BTB}, that will be discussed in Section 4.1.3.

v. Surface dark current (J_{Sh}): Surface dark current arises primarily from trap-assisted generation processes at the surface of the photodiode. During the fabrication process, the semiconductor is typically etched from the top p-region to the bottom n-region to place metal contacts on both sides. This etching exposes the semiconductor to air, creating an abrupt semiconductor-air interface. This interface introduces numerous trap states and trap energy levels, which act as intermediate states within the bandgap. Under reverse bias, these surface traps can capture and release carriers, effectively generating electron–hole pairs and contributing to dark current. Surface passivation techniques are therefore essential to reduce surface dark current, as they help to neutralize or reduce these surface traps and improve photodiode performance. The J_{Sh} equation is,

$$J_{Sh} = \frac{V_a}{R_{sh}} \cdot A \tag{22}$$

Crucial variables: To minimize J_{Sh}, increasing R_{sh} through surface passivation fabrication techniques is essential, as these reduce surface traps and improve surface quality. Although the equation may suggest that reducing device area would lower surface current, in practice, surface current becomes more prominent in small-area devices due to the difference in how surface and bulk areas scale. For instance, while bulk volume scales with πR^2 (where R is the device radius), the surface area scales with $2\pi R$. As device dimensions decrease, the bulk volume reduces more rapidly than the surface area, leading to a higher surface-to-volume ratio. This shift causes surface dark current to dominate. Additionally, surface current can be further reduced by optimizing the electrostatic properties, as discussed in Section 4.3.

4. Elcctrostatic engineering of photodiodes

In the previous section, a photodiode's material-dependent variables and material-independent electrostatic variables—such as charge or doping, thickness, and electric field—and their roles in reducing signal and noise components were discussed. With advancements in semiconductor growth techniques, materials can be engineered to fine-tune the parameters for a high sensitivity and low noise [12–16]. On the other hand, electrostatic engineering, which involves techniques such as barrier structures and field optimization, can be implemented independently of the material. This section will focus on electrostatic engineering to further enhance performance by reducing dark current and other noise sources.

The previous section discusses sensor material-dependent variables and material-independent electrostatic variables—such as charge or doping, thickness, and electric field—and their roles in reducing signal and noise components. Additionally, electrostatics can be engineered using barrier structures and field engineering to reduce dark current.

4.1 Barrier structures

Barrier structures are heterojunction photodiodes engineered to suppress the dark current. Photodiodes can be classified into two types: homojunction and heterojunction photodiodes. Homojunction photodiodes are composed of a single semiconductor material. Heterojunction photodiodes are often made from two or more types of semiconductor materials. Previously, heterostructure materials were chosen with lattice constants close to each other [17]. With recent developments in device fabrication, highly mismatched materials can now be successfully integrated to achieve high-performance sensors, catering to the requirements of advanced camera systems.

The band alignment of the heterostructure is of utmost importance, as it determines the flow of electrons and holes between the materials. The three types of band alignments are straddling gap (type I), staggered gap (type II), and broken gap (type III) heterostructures. Electron affinity, work function, and bandgap determine the type of heterostructure, and the conduction and valence band offset between two materials.

Figure 5 illustrates the equilibrium and non-equilibrium p-doped Type II band alignment of a heterostructure. In non-equilibrium state, the materials are isolated,

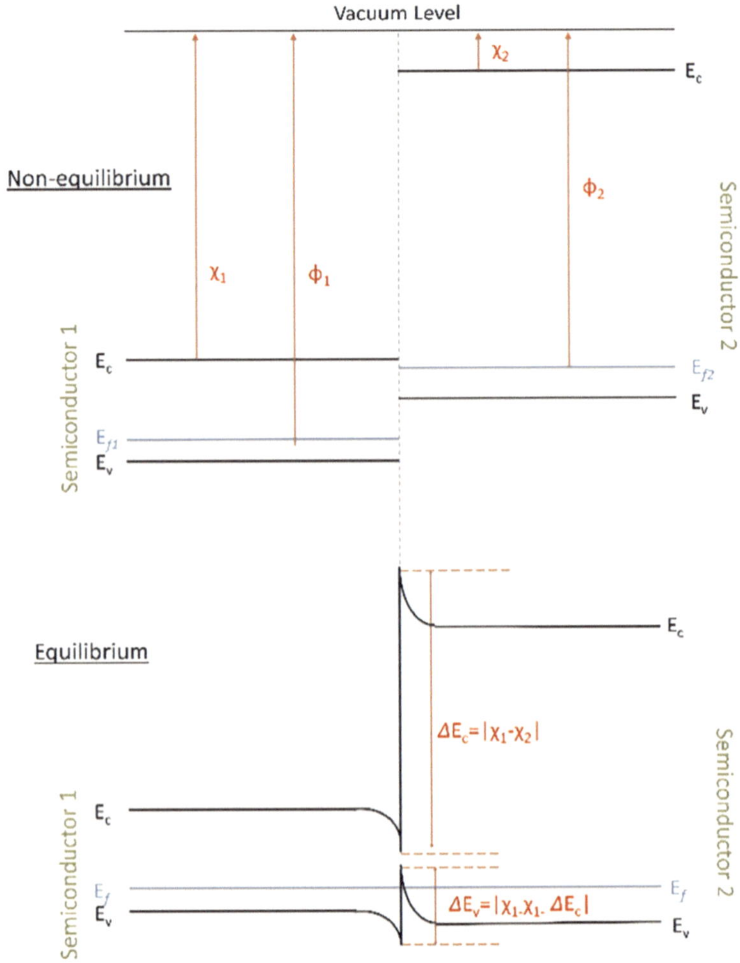

Figure 5.
Non-equilibrium and equilibrium energy band diagram of two p-doped-semiconductor materials indicating their electron affinity and work function (type II heterostructure).

and their energy levels are aligned to the vacuum level. The electron affinity (χ), work function (ϕ), and the bandgaps shown represent the key properties influencing band alignment. In the equilibrium, the materials form a junction, charge transfer occurs at the interface until the Fermi levels align across the system. This alignment results in band bending and the offsets in the conduction and valence bands. The conduction band offset (ΔE_c) and valence band offset (ΔE_v) are shown, highlighting the energy barriers for electron and hole transport.

By carefully engineering the band offsets and selecting appropriate materials, barrier structures can be designed to selectively enhance or block electron and hole transport, enabling the suppression of specific dark current mechanisms.

4.1.1 Barrier structures to reduce diffusion dark current

Figure 6 is the energy band diagram for a homojunction PIN and a heterojunction PBIBN photodiodes at a small reverse bias (V_j). In the PIN structure, the P region represents the p-doped semiconductor, the I region is intrinsic (undoped), and the N region is n-doped. Here, all the regions are photoactive and generate photo carriers. In the PBIBN structure, the B region represents the barrier, which is p-doped on the P side and n-doped on the N side and engineered with a bandgap larger than that of the photoactive material [18]. The barrier material is not photoactive. This band diagram simplified depiction, and actual band bending is more complex in a real device.

In homojunction PIN photodiodes under reverse bias, diffusion dark current arises from the minority carriers in the P and N regions as discussed in Section 3.3. The minority carriers (electrons) diffuse into the depletion region from the P region, while the minority carriers (holes) diffuse into the depletion region from N region. These diffusion currents are driven by the concentration gradients of the carriers. The magnitude of these currents is exponentially related to the intrinsic carrier concentrations in the P and N regions (as given by Eq. (11)), which in turn depend on the material's bandgap. For silicon, with a bandgap of 1.12 eV, diffusion dark current is low but becomes pronounced as the bandgap decreases for materials optimized for infrared sensing.

Heterojunction PBIBN photodiodes are designed to suppress diffusion dark current by introducing barrier regions with a higher bandgap than the photoactive material [18, 19]. On the P side, the barrier creates a large conduction band offset that blocks the diffusion of minority carriers (electrons) from the P region. At the same time, the valence band offset is minimized to ensure that photogenerated holes in the intrinsic region can flow freely toward the P side. High p-type doping in the barrier mitigates any residual valence band offset, further ensuring efficient hole transport while effectively blocking electron diffusion.

Similarly, on the N side, the barrier creates a large valence band offset that blocks the diffusion of minority carriers (holes) from the N region. The conduction band offset is minimized to allow photogenerated electrons in the intrinsic region to flow freely toward the N side. High n-type doping in the barrier mitigates any residual conduction band offset, ensuring efficient electron transport while effectively blocking hole diffusion.

By carefully engineering the barriers on both sides with appropriate band offsets, the PBIBN photodiode significantly reduces diffusion dark current without compromising the transport efficiency of photogenerated carriers. For infrared applications using low-bandgap materials, the high diffusion dark current makes PBIBN structures an ideal choice for reduced noise performance [20].

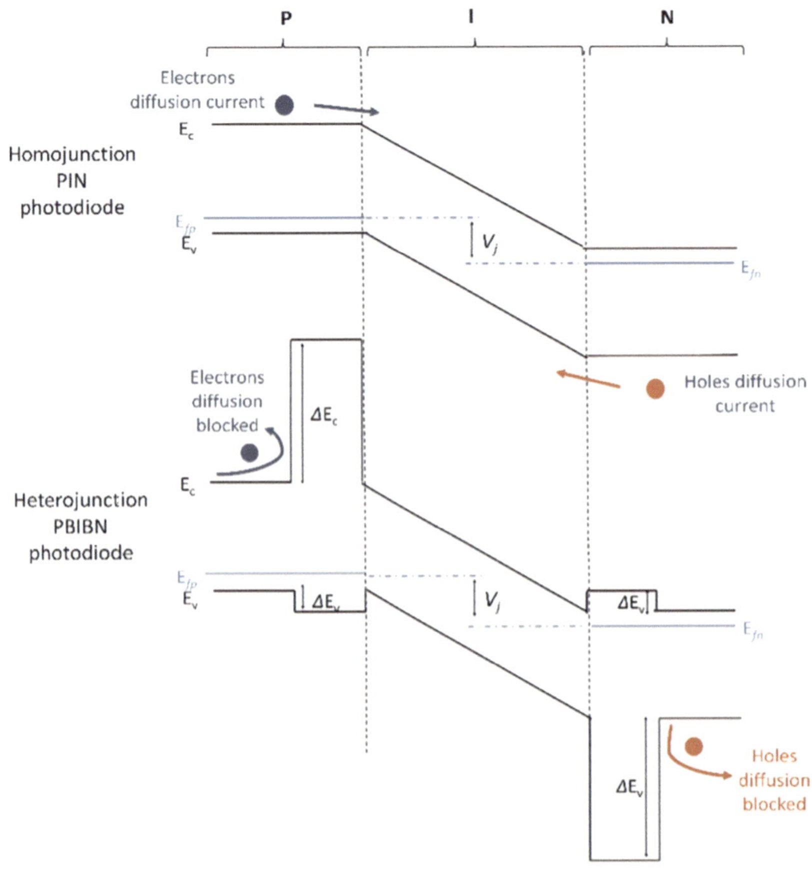

Figure 6.
Non-equilibrium energy band diagram of PIN homojunction and PBIBN heterojunction photodiodes, illustrating the flow of diffusion dark current.

4.1.2 Barrier structures to reduce generation-recombination dark current

Figure 7 illustrates the energy band diagrams of PIN homojunction and NBN heterojunction photodiodes under a small reverse bias voltage. In the PIN structure, all regions are photoactive, whereas, in the NBN structure, the B region serves as a barrier with a larger bandgap than the photoactive material. In a PIN photodiode, the active material's bandgap is selected based on the target wavelength range, with narrower bandgaps often used for infrared sensing. As discussed in Section 3.3, the G-R dark current is susceptible to lower bandgap (due to high intrinsic carrier concentration), and a depletion region with traps in the mid-bandgap that facilitate carrier transitions from the valence band to the conduction band. All these conditions are prominent in the narrow bandgap PIN structure. This issue becomes more pronounced as the target wavelength shifts from VIS to MWIR.

In contrast, the NBN heterojunction photodiode significantly suppresses G-R dark current by introducing a wide bandgap barrier in the B region, which occupies a substantial portion of the depletion region [21]. The wide bandgap in the B region reduces and minimizes the density of thermally generated carriers. Additionally, mid-gap trap states in the B region are less likely to facilitate carrier transitions due to the increased difficulty of thermal excitation across the larger bandgap. As a result,

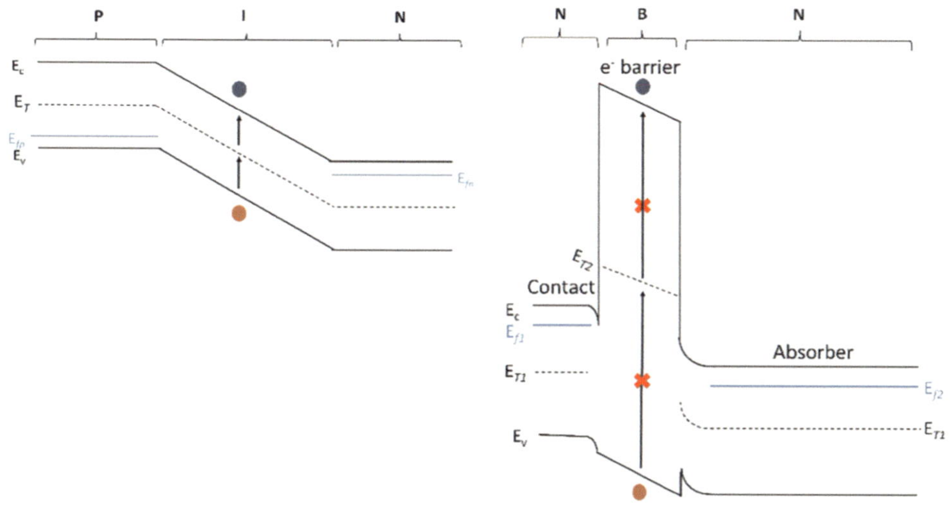

Figure 7.
Non-equilibrium energy band diagram of PIN homojunction and NBN barrier heterojunction photodiodes, illustrating the flow of GR current.

the G-R current is considerably lower in the NBN structure compared to the PIN photodiode.

A small portion of the depletion region in the NBN photodiode remains in the narrow bandgap N region, contributing to G-R dark current. This can be minimized by optimizing doping profiles and barrier thickness to confine the electric field and potential drop primarily within the wide bandgap barrier.

By shifting much of the depletion region into the wide bandgap barrier with low intrinsic carrier concentration, the impact of carrier transitions from the valance band to the conduction band via the mid-gap trap states is minimized, suppressing the G-R current and. By shifting much of the depletion region into the wide bandgap barrier with low intrinsic carrier concentration, the impact of carrier transitions from the valance band to the conduction band via the mid-gap trap states is minimized, suppressing the G-R current. The suppression of the noise.

4.1.3 Barrier structures to reduce tunneling current dark current

Figure 8 illustrates the energy band diagrams of PIN homojunction and NBN heterojunction photodiodes under a small reverse bias voltage, showing the flow of trap-assisted tunneling (TAT) and band-to-band tunneling (BTB) dark currents.

In the PIN structure, the high electric field in the narrow bandgap I (intrinsic) region facilitates both TAT and BTB tunneling, contributing significantly to dark current. TAT occurs via mid-bandgap trap states, while BTB allows direct tunneling from the valence to the conduction band, both exacerbated by the narrow bandgap and strong field as discussed in Section 3.3 [22].

In the NBN structure, the wide bandgap barrier (B) region effectively blocks tunneling currents. The increased bandgap heightens the energy separation, and the thicker barrier elongates the tunneling path, reducing tunneling [21]. However, some high fields remain in the photoactive N region, which can lead to residual tunneling currents.

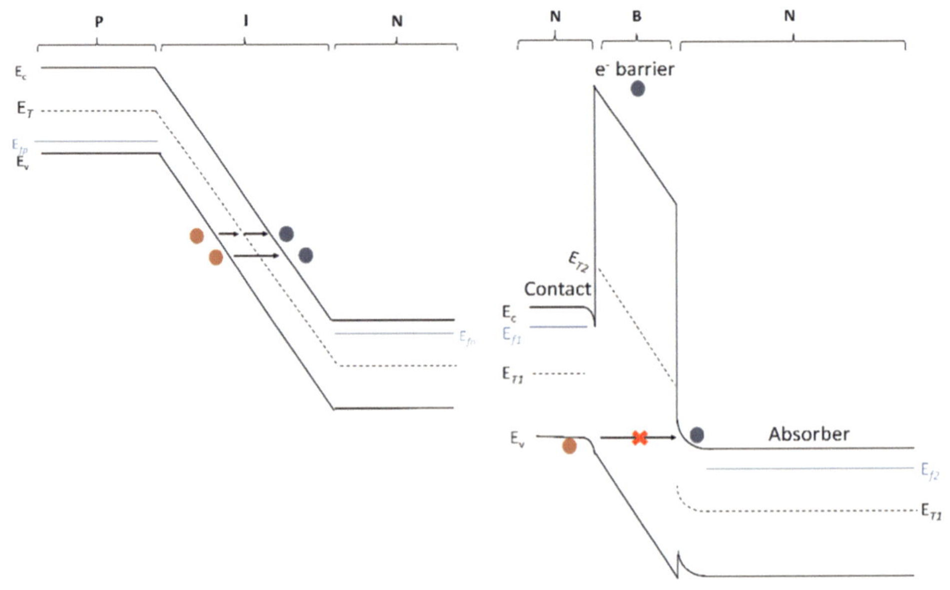

Figure 8.
Non-equilibrium energy band diagram of PIN homojunction and NBN barrier heterojunction photodiodes, illustrating the flow of TAT and BTB dark current.

Careful electrostatic design, such as optimizing doping profiles and barrier thickness, minimizes the electric field in the photoactive region while confining it to the barrier. This significantly reduces tunneling dark current, improving noise performance.

4.1.4 Barrier structures to reduce surface current

Surface dark current arises from the thermally generated carriers at the semiconductor surface due to surface defects as discussed in Section 3.3. These defects often lead to an excess of electrons or holes at the surface, even in undoped semiconductors. For n-type surfaces with a high density of surface electrons, this can contribute significantly to surface dark current.

The NBN barrier structure effectively suppresses surface dark current with a wide bandgap barrier [21]. This barrier creates an energy offset that prevents surface electrons from entering the conduction band, thereby blocking their contribution to dark current. As shown in **Figure 9**, the conduction band offset in the barrier region acts as an electron-blocking layer, significantly reducing the flow of surface electrons into the absorber region, reducing the noise.

4.2 Field engineering to suppress surface dark current

Field engineering refers to strategies designed to optimize the electric field distribution within photodiodes to enhance the performance metrics. Though there are several examples of field engineering [8, 14, 23–26], this section focuses on suppressing surface dark current.

The surface dark current can be either reduced by increasing the surface resistance (R_{sh}) through passivation materials, or by reducing the potential drop (or electric field) at the surface, as given by Eq. (22). Field engineering strategies, such as

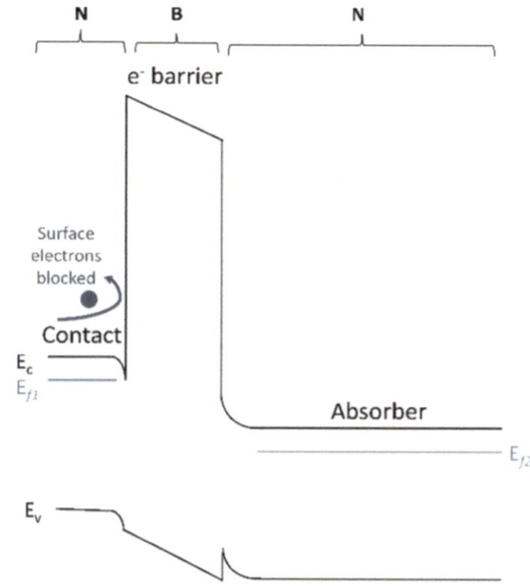

Figure 9.
Non-equilibrium energy band diagram of a NBN barrier heterojunction photodiode, illustrating the suppression of surface dark current.

employing a double mesa structure, are effective in suppressing surface dark current by reducing the electric field at the surface [9–11, 20]. Photodiodes generally have mesa structures, where the semiconductor is etched from the top p-region to the bottom n-region during fabrication to create an isolated region for device operation and to allow the placement of metal contacts on both sides. This etching process creates surface dark current as discussed in Section 3.3.

Double mesa structures go a step further by introducing an additional etching step to create a second, smaller mesa region within the device. This geometry confines the depletion region away from the surface and reduces the availability of negative charges in the P region near the surface that would otherwise image the positive charges in the N region. As shown in **Figure 10**, the electric field in single mesa photodiodes extends uniformly across the bulk and the surface due to the alignment of charges in the P and N regions. This high surface field promotes the surface dark current. In double mesa photodiodes, however, the absence of negative charges in the

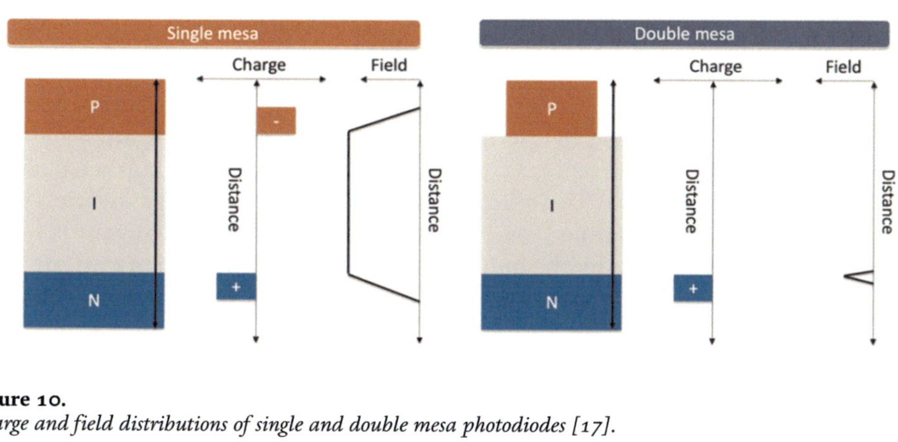

Figure 10.
Charge and field distributions of single and double mesa photodiodes [17].

P region near the surface drastically reduces the electric field at the surface. This lowers the potential drop at the surface lowering the surface dark current. In addition to double mesa structures, it is also recommended to passivate the surface to reduce any residue surface dark current [8].

5. Photodiodes in cameras for earth observation

Earth observation satellites use cameras operating across visible (VIS), short-wave infrared (SWIR), and mid-wave infrared (MWIR) spectral ranges, each tailored to specific applications. While cameras can be categorized into area scanners and line scanners, earth observation satellites typically employ line scanners (push-broom scanners) to take advantage of the relative motion between the satellite and the earth. Line scanners use a one-dimensional sensor array to scan successive lines of the Earth's surface as the satellite moves. These lines are then combined to form a two-dimensional image, making them highly efficient for generating large-scale maps with minimal redundancy. The next sections focus is on the qualitative design of the photodiodes in line scanner cameras for specific applications in VIS, SWIR, and MWIR spectral ranges.

5.1 Photodiodes in VIS cameras for high-resolution imaging

VIS cameras (400–700 nm) play a vital role in earth observation for applications such as urban mapping, infrastructure monitoring, and disaster response.

5.1.1 Material choice

For such applications, silicon is the ideal detector material due to its unique properties [27]:

i. QE: With a bandgap of 1.12 eV, silicon can effectively cover the entire visible spectrum (400–700 nm), ensuring high QE across this range.

ii. Bandwidth: Silicon's high carrier mobility contributes to increased bandwidth, enabling fast charge transfer and better performance in applications requiring high-speed imaging.

iii. Dark current: Over decades of technological advancements, silicon's material quality has been refined to minimize structural defects, significantly reducing G-R and TAT dark currents, as discussed in Section 3.3. The surface dark current is suppressed with a silicon dioxide or silicon nitride passivation layer, which is already fine-tuned over generations. The BTB dark current is minimized by increasing the depletion width, and the diffusion current remains low at room temperature due to silicon's relatively small intrinsic carrier concentration (Eq. (21)).

These characteristics make silicon a high-performance photodetector material for VIS cameras.

5.1.2 Challenge

Achieving high-resolution imaging for t4.3hese applications is critical to capturing fine surface details of the scene. However, high resolution often necessitates smaller pixel sizes, which can negatively affect the SNR. The noise associated with smaller pixels is predominantly influenced by surface dark current. However, decades of advancements in silicon interface technology and the development of effective passivation materials have significantly mitigated this noise. The signal is affected as the smaller pixels collect fewer photons. While increasing the telescope aperture or modifying other optical parameters can improve photon collection, these approaches are often impractical due to their cost and complexity.

5.1.3 Electrostatic solution

Time delay integration (TDI) in line scanner cameras provides an effective alternative to achieve high resolution without significantly compromising SNR [28]. As the satellite moves, the same scene passes through successive rows of pixels in the detector array. TDI cameras exploit this principle by synchronizing the motion of charge carriers within the photodetector array with the relative motion of the scene. As the scene shifts from one row of pixels to the next due to the satellite's motion, the generated charge in each pixel is electronically shifted to the corresponding pixel in the next row, maintaining alignment with the scene. This synchronized charge transfer ensures that photons collected over multiple exposures are combined into a single signal. By integrating the charge collected in this manner, TDI significantly enhances the signal strength, offsetting the limitations imposed by smaller pixels and improving SNR [28].

TDI line scanners often employ charge-coupled devices (CCDs), which consist of an array of simple P-N junction photodiodes made from silicon, with a silicon dioxide layer on top serving as an insulator and gate structure, as shown in **Figure 11** [29, 30]. When a reverse bias is applied through the gate, a depletion region is formed at the P-N junction, that store photogenerated charge when a light is incident. By adjusting

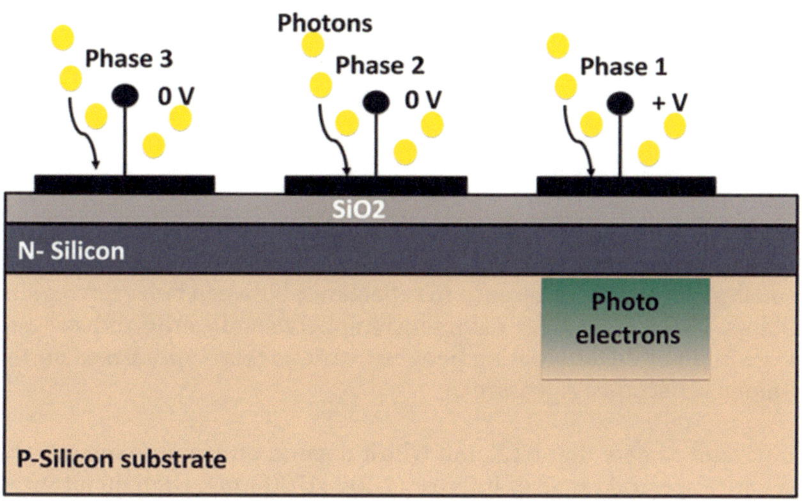

Figure 11.
CCDs array pixel structure.

gate voltages in phases, the charge is shifted efficiently across the array, synchronizing with the satellite's motion. This ensures that the charge collected from the same scene by successive rows in the array is integrated, effectively increasing the signal strength.

The number of rows in the array, or TDI stages, can be increased to achieve the desired SNR while maintaining high resolution [30]. For an array with N rows, the signal from the scene becomes N times the signal from one row as the contributions from each row are added together. However, noise—primarily shot noise—also increases, but only as the square root of N. This means that the SNR improves proportionally to the square root of N, allowing for enhanced sensitivity without compromising the resolution.

By leveraging the exceptional material properties of silicon and the electrostatic engineering using the charge transfer mechanisms of TDI-enabled CCDs, VIS cameras provide a robust solution for high-resolution imaging in earth observation, enabling precise data collection for urban mapping, infrastructure monitoring, and disaster response with superior signal-to-noise performance.

5.2 SWIR photodiodes in cameras for precision farming

Crop health monitoring and crop type differentiation are critical applications in precision farming, enabling optimized resource allocation, early stress detection, and efficient harvesting [1]. SWIR photodiodes in line scanner cameras are particularly effective for these applications due to their ability to detect subtle changes in reflectance in the SWIR spectrum. Reflectance, which represents the fraction of light reflected off the vegetation canopy, directly influences the normalized light intensity (irradiance) reaching the detector. These changes are especially significant in the 1000–1700 nm range, where variations in leaf water content and biochemical composition alter the spectral signature. In the SWIR spectral range, the reflectance of sunlight off terrestrial objects dominates over their thermal emittance, making reflectance-based imaging the primary mechanism for precision farming.

To ensure accurate crop health monitoring and crop type differentiation, two key metrics must be considered in the design of SWIR photodiodes:

i. Noise equivalent reflectance (NER): NER represents the smallest absolute reflectance value that the sensor can detect above the noise floor, where the signal-to-noise ratio (SNR) is equal to 1. It reflects the sensor's ability to detect low reflectance levels, such as faint signals from stressed vegetation or sparse canopies.

ii. Noise equivalent differential reflectance (NEdR): NEdR measures the smallest detectable difference in reflectance between two crop regions or time points. This is crucial for distinguishing between healthy and stressed vegetation or differentiating between various crop types based on their unique reflectance signatures.

It is important to note that NER and NEdR depend on maintaining a high SNR across the entire spectral range of interest (1000–1700 nm). High quantum efficiency (QE) and low noise are essential for achieving this SNR, ensuring accurate detection and differentiation in precision farming applications.

5.2.1 Material choice

For SWIR applications, HgCdTe (MCT) and InGaAs are widely used. However, MCT has several issues that include complex growth technique and lower yields, susceptible to aging effects like defects, high dark current at room temperature and a need for expensive cooling [31, 32]. On the other hand, InGaAs technology is more matured, reliable, and does not require cooling at room temperature [25]. Its properties include:

i. QE: With a bandgap of 0.732 eV, silicon can effectively cover the entire spectral range of interest (1000–1700 nm), ensuring high QE across this range.

ii. Bandwidth: Thanks to its high mobility, the bandwidth is high.

iii. Dark current: Like silicon, InGaAs has benefited from decades of technological advancements, leading to significant improvements in material quality. These advancements have minimized structural defects, thereby reducing G-R and TAT dark currents, as discussed in Section 3.3. The BTB dark current is minimized by increasing the depletion width as discussed in Section 3.3. The BTB tunneling threshold for the InGaAs is about 180 kV/cm [33]. At operating voltage, the thickness of the depletion width can be adjusted so the field drop is lower than the tunneling threshold.

5.2.2 Challenges

InGaAs has a high diffusion current as the intrinsic carrier concentration is high because of a lower bandgap as shown in Eq. (11). In addition, the surface is conductive, leading to a high surface dark current.

5.2.3 Electrostatic solutions

The diffusion dark current can be suppressed using the PBIBN barrier architecture discussed in **Figure 6** of Section 4.1.1. Here, the barrier material can be InAlAs, with a wide bandgap of 1.47 eV, lattice matched to the InGaAs [34]. The surface dark current can be reduced by employing double mesa structures discussed in **Figure 10** of Section 4.3. SU-8, Al_2O_3, SiO_2 or Si_3N_4 can be used as passivation to further reduce the surface dark current.

By combining high-performance materials like InGaAs with advanced electrostatic engineering techniques, SWIR photodiodes in line scanners EO cameras provide a powerful solution for precision farming, enabling accurate crop health monitoring and differentiation, ultimately supporting sustainable agricultural practices.

5.3 MWIR photodiodes in cameras for environmental monitoring

MWIR cameras play a crucial role in environmental monitoring applications, such as detecting greenhouse gases, thermal anomalies, and surface temperature variations [2, 35, 36]. These applications often require sensors capable of accurately detecting and distinguishing temperature differences within the 3–5.5 μm range, corresponding to the thermal emission of many atmospheric gases and surface heat patterns. In the

MWIR spectral range, the thermal emittance off the terrestrial objects dominates over the reflectance of sunlight, making thermal emittance-based imaging the primary mechanism for environmental monitoring.

To ensure accurate temperature-based measurements, two key performance metrics must be considered when designing MWIR photodiodes:

i. Noise equivalent temperature (NET): NET represents the smallest absolute temperature difference the sensor can resolve above the noise floor, where the signal-to-noise ratio (SNR) equals 1. This metric reflects the photodiode's sensitivity to weak thermal signals, such as those emitted by low-concentration greenhouse gases or faint thermal variations in the environment. A lower NET helps detect the lowest thermal signatures.

ii. Noise equivalent differential temperature (NEdT): NEdT measures the smallest detectable temperature difference between two regions or time points. This is crucial for environmental monitoring tasks such as identifying localized thermal anomalies, tracking changes in surface temperature over time, or detecting small variations in the thermal emission of greenhouse gases. A low NEdT ensures that the sensor can differentiate between minor temperature gradients effectively.

Both NET and NEdT are directly tied to the SNR of the photodiode, and it is important to have a high SNR across the entire spectral range of interest (3–5.5 μm).

5.3.1 Material choice

For MWIR applications, MCT is considered a high-performance photodiode. However, the need for cryogenic cooling at MWIR spectral range, and a lower yield are reducing its cost-effectiveness. InGaAs/InAsSb superlattice (SL) is one of the recommended detector materials [37]. It only requires thermoelectric colling as opposed to complex cryogenic cooling of MCTs and has a better yield. The InGaAs/InAsSb SL properties include [37]:

i. QE: With a bandgap of 0.215 eV, the InGaAs/InAsSb can effectively cover the entire MWIR range of interest, ensuring high QE across the range.

ii. Bandwidth: Thanks to its high mobility, the bandwidth is high.

iii. Dark current: Though the dark current noise is high for InGaAs/InAsSb SL, barrier structures with AlGaAsSb can be used to suppress it.

5.3.2 Challenges

All III-V based MWIR photodetectors including InGaAs/InAsSb SL, suffer from a high dark current at room temperature.

5.3.3 Electrostatic solutions

All dark current mechanisms in the InGaAs/InAsSb superlattice (SL) can be effectively suppressed using an NBN barrier structure, as discussed in Section 4.1. The barrier material can be the lattice matched wide bandgap AlGaAsSb. G-R dark current is suppressed by confining most of the depletion region within the wide bandgap barrier, as illustrated in **Figure 7**. BTB and TAT are blocked by the high tunneling threshold of the AlGaAsSb barrier, as shown in **Figure 8**. The barrier's tunneling threshold is 600 kV/cm, effectively preventing tunneling currents [38]. Surface dark current is minimized by blocking the n-type surface carriers with the barrier, as discussed in **Figure 9**.

For efficient carrier transport, the valence band offset between the barrier and the InGaAs/InAsSb SL layer must be minimal. The 0-eV band offset between the InGaAs/InAsSb SL and the AlGaAsSb, ensures unhindered movement of photogenerated holes from the active layer to the P-side contact, improving the QE [28].

By leveraging the high-performance InGaAs/InAsSb superlattice material and electrostatic engineering techniques, MWIR photodiodes in line scanner EO cameras offer a robust solution for environmental monitoring, enabling precise detection of greenhouse gases, thermal anomalies, and surface temperature variations, ultimately contributing to more effective environmental management and sustainability efforts.

6. Conclusions

The integration of advanced electrostatic engineering principles with innovative materials has revolutionized the design of photodiodes for earth observation applications. By tailoring photodetectors to specific spectral ranges, from VIS to SWIR to MWIR, the diverse demands of urban monitoring, precision farming, and environmental monitoring are met. The are met. The strategic use of several electrostatic engineering techniques like the time delay integration, the barrier structures, and the field engineering, optimizes quantum efficiency, minimizes dark currents, and enhances signal-to-noise ratios (SNR) can be optimized across all applications. Through the strategic use of several electrostatic engineering techniques like the time delay integration, the barrier structures, and the field engineering, optimizes quantum efficiency, minimizes dark currents, and enhances signal-to-noise ratios (SNR) can be optimized across all applications.

These developments underscore the importance of material selection, such as silicon, InGaAs, and InGaAs/InAsSb superlattices, and highlight the role of electrostatic solutions in addressing challenges like high dark current and low SNR. Collectively, these innovations provide the foundation for sustainable and effective earth observation technologies, enabling precision, efficiency, and long-term impact in critical global applications.

Author details

Sri Harsha Kodati[1*] and Manisha Muduli[2]

1 Planet Labs PBC, San Francisco, California, United States of America

2 The Ohio State University, Columbus, Ohio, United States of America

*Address all correspondence to: sriharshakodati@gmail.com

References

[1] Pande CB, Moharir KN. Application of hyperspectral remote sensing role in precision farming and sustainable agriculture under climate change: A review. In: Climate Change Impacts on Natural Resources, Ecosystems and Agricultural Systems. Springer Climate. Cham, Switzerland: Springer; 2023. pp. 503-520

[2] LeVan PD, Sakoglu U. Infrared sensing technologies assisting environmental monitoring. Proc. SPIE 11503. In: Infrared Sensors, Devices, and Applications X. Vol. 11503. Bellingham, Washington, USA: SPIE; 2020. pp. 38-49

[3] Rogalski A, Martyniuk P, Kopytko M. Challenges of small-pixel infrared detectors: A review. Reports on Progress in Physics. 2016;**79**(4):046501

[4] Delmas M, Rodriguez J-B, Christol P. Electrical modeling of InAs/GaSb superlattice mid-wavelength infrared PIN photodiode to analyze experimental dark current characteristics. Journal of Applied Physics. 2014;**116**(11):114505

[5] Jung H, Lee S, Liu Y, Jin X, David J, Krishna S. High electric field characteristics of GaAsSb photodiodes on InP substrates. Applied Physics Letters. 2023;**122**(22):221102

[6] Braga OM, Delfino CA, Kawabata RM, Pinto LD, Vieira GS, Pires MP, et al. Surface passivation of InGaAs/InP pin photodiodes using epitaxial regrowth of InP. IEEE Sensors Journal. 2020;**20**(16):9234-9244

[7] Muduli M, Xia Y, Lee S, Gajowski N, Chae C, Rajan S, et al. Active interface characteristics of heterogeneously integrated GaAsSb/Si photodiodes. Applied Physics Letters. 2024;**125**(17):171102

[8] Muduli M, Schwartz M, Gajowski N, Lee S, Krishna S. Investigation of Zn diffusion in 2-micron InGaAs/GaAsSb superlattice planar diodes using atomic layer deposition of ZnO. Proc. SPIE 12534. In: Infrared Technology and Applications XLIX. Vol. 12534. SPIE; 2023. pp. 45-50

[9] Fink DR, Lee S, Kodati SH, Dahiya V, Ronningen TJ, Winslow M, et al. Determination of background doping type in type-II superlattice using capacitance-voltage measurements with double mesa structure. Proc. SPIE 11407. In: Infrared Technology and Applications XLVI. Vol. 11407. Bellingham, Washington, USA: SPIE; 2020. pp. 24-34

[10] Schwartz M, Kodati S, Lee S, Jung H, Chen D, Grein C, et al. Determination of the background doping polarity for unintentionally doped AlGaAsSb and AlInAsSb avalanche photodiodes on InP substrates. AIP Advances. 2022;**12**(9):6

[11] Schwartz M, Kodati SH, Lee S, Jung H, Chen D, Grein C, et al. Determination of doping polarity of unintentionally doped antimonide avalanche photodiodes on InP substrate. Proc. SPIE 12107. In: Infrared Technology and Applications XLVIII. Vol. 12107. Bellingham, Washington, USA: SPIE; 2022. pp. 105-112

[12] Nguyen B-M, Hoffman D, Delaunay P-Y, Razeghi M. Dark current suppression in type II InAs/GaSb superlattice long wavelength infrared photodiodes with m-structure barrier. Applied Physics Letters. 2007;**91**(16):163511

[13] Lee S, Kodati S, Guo B, Jones A, Schwartz M, Winslow M, et al. Low noise Al0. 85Ga0. 15As0.56Sb0. 44

avalanche photodiodes on InP substrates. Applied Physics Letters. 2021;**118**(8): 081102

[14] Kodati S, Lee S, Guo B, Jones A, Schwartz M, Winslow M, et al. AlInAsSb avalanche photodiodes on InP substrates. Applied Physics Letters. 2021;**118**(9): 091103

[15] Lee S, Winslow M, Grein C, Kodati S, Jones A, Fink D, et al. Engineering of impact ionization characteristics in In0. 53Ga0. 47As/Al0. 48In0. 52As superlattice avalanche photodiodes on InP substrate. Scientific Reports. 2020;**10**(1):16735

[16] Lee S, Guo B, Kodati S, Jung H, Schwartz M, Jones A, et al. Random alloy thick AlGaAsSb avalanche photodiodes on InP substrates. Applied Physics Letters. 2022;**120**(7):071102

[17] Davies JH. The Physics of Low-Dimensional Semiconductors: An Introduction. Cambridge, United Kingdom: Cambridge University Press; 1998

[18] Gautam N, Myers S, Barve AV, Klein B, Smith E, Rhiger D, et al. Band engineered HOT midwave infrared detectors based on type-II InAs/GaSb strained layer superlattices. Infrared Physics & Technology. 2013;**59**: 72-77

[19] Gautam N, Kim H, Kutty M, Plis E, Dawson L, Krishna S. Performance improvement of longwave infrared photodetector based on type-II InAs/GaSb superlattices using unipolar current blocking layers. Applied Physics Letters. 2010;**96**(23):231111

[20] Agrawal GP. Fiber-Optic Communication Systems. Hoboken, New Jersey, USA: John Wiley & Sons; 2012

[21] Uzgur F. Barrier engineering for high-performance nBn infrared photodetectors [master's thesis]. Ankara, Turkey: Middle East Technical University; 2020.7

[22] Kopytko M, Rogalski A. New insights into the ultimate performance of HgCdTe photodiodes. Sensors and Actuators A: Physical. 2022;**339**:113511

[23] Kodati SH. Low-noise Antimonide-Based Avalanche Photodiodes on InP Substrates. PhD thesis. Columbus, Ohio, USA: The Ohio State University; 2022

[24] Krishna S, Kodati SH, Ronningen T, Lee S. Lateral Interband Type II Engineered (LITE) Detector. US Patent 11,508,869. 2022

[25] Krishna S, Lee S, Kodati SH, Schwartz M, Jung H, Ronningen TJ, et al. Linear mode avalanche photodiodes with antimonide multipliers on InP substrates. IEEE Journal of Quantum Electronics. 2022;**58**(4):1-7

[26] Guo B, Schwartz M, Kodati SH, McNicholas KM, Jung H, Lee S, et al. InGaAs/AlInAsSb avalanche photodiodes with low noise and strong temperature stability. APL Photonics. 2023;**8**(11):111101

[27] Fang Z, Zhao CZ. Recent progress in silicon photonics: A review. International Scholarly Research Notices. 2012; **2012**(1):428690

[28] Li Y-H, WangX-D. Compressive imaging beyond the sensor's physical resolution via coded exposure combined with time-delay integration. Optics and Lasers in Engineering. 2023;**164**:107491

[29] Boyle WS, Smith GE. Charge coupled semiconductor devices. Bell System Technical Journal. 1970;**49**(4): 587-593

[30] Milton AF. Charge transfer devices for infrared imaging. In: Optical and Infrared Detectors. Springer; 2005. pp. 197-228

[31] Yuan H, Zhang J, Kim J, Bond D, Laquindanum J, Kimchi J, et al. Recent progress in extended wavelength InGaAs photodetectors and comparison with SWIR HgCdTe photodetectors. Proc. SPIE 11129. In: Infrared Sensors, Devices, and Applications IX. Vol. 111290E. Bellingham, Washington, USA: SPIE; 2019. pp. 97-106

[32] Rogalski A, Martyniuk P. InAs/GaInSb superlattices as a promising material system for third generation infrared detectors. Infrared Physics & Technology. 2006;**48**(1):39-52.8

[33] Cao S, Zhao Y, ur Rehman S, Feng S, Zuo Y, Li C, et al. Theoretical studies on InGaAs/InAlAs SAGCM avalanche photodiodes. Nanoscale Research Letters. 2018;**13**:1-15

[34] Goh Y, Massey D, Marshall A, Ng J, Tan C, Ng W, et al. Avalanche multiplication in InAlAs. IEEE Transactions on Electron Devices. 2006;**54**(1):11-16

[35] Zhong Y, Wang X, Wang S, Zhang L. Advances in spaceborne hyperspectral remote sensing in China. Geo-Spatial Information Science. 2021;**24**(1):95-120

[36] Hollenbeck D, Zulevic D, Chen Y. Advanced leak detection and quantification of methane emissions using sUAS. Drones. 2021;**5**(4):117

[37] Newell A, Logan J, Carrasco R, Alsaad Z, Hains C, Duran J, et al. Effects of doping and minority carrier lifetime on mid-wave infrared InGaAs/InAsSb superlattice nBn detector performance. Applied Physics Letters. 2023;**122**(17):171101

[38] Lee S, Jin X, Jung H, Lewis H, Liu Y, Guo B, et al. High gain, low noise 1550 nm GaAsSb/AlGaAsSb avalanche photodiodes. Optica. 2023;**10**(2):147-154

Chapter 6

Functional Perspectives of Endogenous Electric Fields in Humans and Rodents: A Viewpoint on Ephaptic Physiology

Toshikazu Shinba

Abstract

The present chapter summarizes the functional perspectives of endogenous static and extremely low-frequency electric fields generated in the body of humans and rodents and discusses the possibility of ephaptic interaction. The electric field recorded on the head surface (electroencephalogram) exhibits μV-range amplitude, which is not significantly affected by the brain size. The surface potential is small, but the electric field inside the brain is regionally localized and exhibits an intensity of around 1 V/m. Electric fields with a similar intensity are present in the muscle and skin. These observations suggest that the endogenous electric field shows complex patterns of distribution inside the body. Furthermore, endogenous electric field profiles are influenced by externally applied electric field. Event-related field potentials in the human and rat brain are altered by concomitant static electric field treatment, and externally applied extremely low-frequency electric field induces an increase in electroencephalogram power. These findings would support the presence of ephaptic interaction, which may be related to the modulation of physiological function. Although further research is necessary, an endogenous electric field can serve as a form of interactive information to integrate physiological activities.

Keywords: endogenous electric field, exogenous electric field, electroencephalogram, local electric field, static electric field, extremely low electric field, ephaptic interaction

1. Introduction

The electric field is generated in the space by substances that are electrically charged and are ubiquitously present on the earth. Externally, both natural and artificial conditions create static and extremely low-frequency electric fields, in addition to higher-frequency electromagnetic fields. Electric and electromagnetic fields of various frequencies affect biological functions differently depending on their frequency ranges [1–4].

Electric fields are also generated endogenously in various parts of the body in the humans and animals including their central and peripheral structures. They are recorded as electroencephalogram (EEG), electrocardiogram (ECG), electromyogram (EMG), skin potential, etc., and are used to understand the functions and dysfunctions of the individual organs. Although these electric fields have been studied and utilized clinically for diagnostic and pathophysiological assessments [5], they are mostly considered as epiphenomena.

Electric fields are generated by various structures containing cells producing transmembrane potentials and spatial potential gradients, but are not considered to convey information affecting the functions of the structures or integrating them. Conventionally, the information inside the body is transferred through nerve fibers, chemical synapses, electrical gap junctions, hormones, and other molecules [6]. In addition to these well-studied components of information processing, electric field has been presented as a candidate for a new type of information to be utilized for prompt cell-to-cell and organ-to-organ interaction and integration, and is called ephaptic coupling [7, 8]. In this theory, the electric field is not an epiphenomenon but is functionally active, containing significant information.

The exact nature of ephaptic coupling is not well clarified. Arguments against ephaptic coupling are based on the low intensity of electric fields. The present chapter describes the data that the electric fields recorded in humans and rodents are large enough to be functionally meaningful, supporting the ephaptic interaction and integration. Discussion includes the surface vs. local electric field and the relation between externally-applied and endogenous electric fields. The data on both static and extremely low-frequency electric fields are presented.

2. Electric fields in the brain

2.1 Amplitude of electroencephalogram (EEG) and brain size

EEG is an endogenous extremely low-frequency electric field generated endogenously in the brain, and is recorded between two electrodes placed at different sites on the head. Clinically, the ear lobe is often used as the indifferent site. The electrode on the head detects the electric field at that site and the difference between two electrodes is calculated as EEG.

In psychophysiological studies, various EEG parameters are used for assessing brain functions and psychological states, and for evaluation of neuropsychiatric disorders including epilepsy, depression, schizophrenia, and dementia [5]. EEG is essential for the diagnosis of epilepsy showing characteristic EEG waveforms including spiky wave of several millisecond duration. Spontaneous fluctuation with the frequency ranging from DC to 30 Hz is also used to evaluate the level of arousal [9] and is employed to analyze the symptoms of dementia and delirium.

Figure 1 shows three types of EEG in normal, demented, and delirious human subjects when they are sitting on a chair with their eyes closed. In both normal and demented subjects, basic wave rhythm is present with the amplitude of μV-order. The wave frequency in this demented subject is 6 Hz and is slower than that in the normal subject (8 Hz). This delirious subject shows EEG fluctuation of lower voltage without clear rhythm formation. Demented subjects exhibit cognitive disturbances and

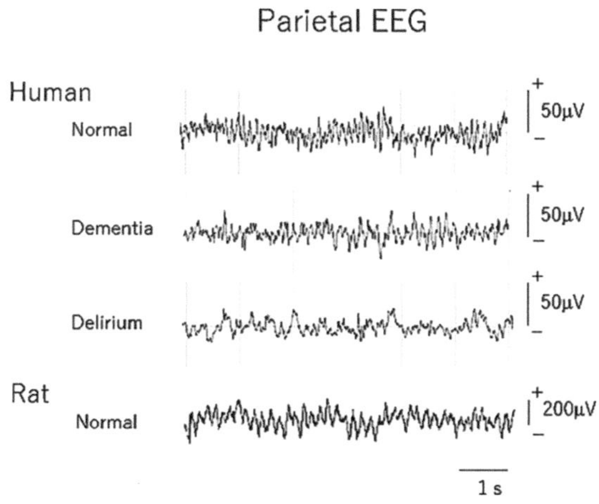

Figure 1.
Electroencephalogram (EEG) of normal, dementia, and delirium states in human subjects with the eyes closed and of normal awake state in a behaving rat. EEG was recorded using an Ag/AgCl electrode at the parietal area with the reference electrode set on the eal lobe in the human subjects and on the cerebellum surface in the rat.

delirious subjects show confusion with a lowered arousal level. EEG is thus related to changes in psychophysiological processes but exhibits small amplitude at the human head surface, being considered an epiphenomenon.

Figure 1 also shows EEG measured on the cortex in a normal awake rat. The frequency and the amplitude of the basic wave are 6 Hz and 200 μV in this recording. The frequency is lower than and the amplitude is larger than the normal human recording (8 Hz, 50 μV). The lower frequency in rat may be due to its generation mainly in the hippocampus, which is situated near the rat brain surface. Human hippocampus is situated in the deeper parts of the brain and the generated field potential is not readily detected from the surface.

Figure 2 shows the contingent negative variation (CNV) measured at the parietal area of the head in a normal human subject. CNV is an endogenous static field recorded on the brain surface during a sustained psychological condition, including attention, preparation, and expectation [10]. CNV is obtained during a delay period, and neuronal correlates are found in the cortex, including frontal and sensory cortices [11, 12]. In **Figure 2**, CNV at the parietal area of the head in a human subject is observed during a Go/NoGo task in which the subject is required to wait for 3 s in response to a Go sign on a display and to press a button as quickly as possible when the imperative sign is turned up. The amplitude of CNV is sustained at 15–20 μV in this subject following the initial event-related potentials, P1 and P3. No negative shift is observed in response to the NoGo sign. It is interesting to see that P1 and P3 components of event-related potentials are clearly present for the NoGo sign, indicating that the neural processes after the Go and NoGo signs are equal until this point of field potential generation.

Figure 3 exhibits CNV recorded in a mouse trained for a discriminative conditioning paradigm using two tones with a delay [12]. Foot electrical stimulation is used as an unconditioned stimulus. CNV appears during the delay period only after the tone stimulation paired with foot stimulation from the third day of training. The amplitude is about 100 μV, and CNV is sustained until the foot stimulation is delivered.

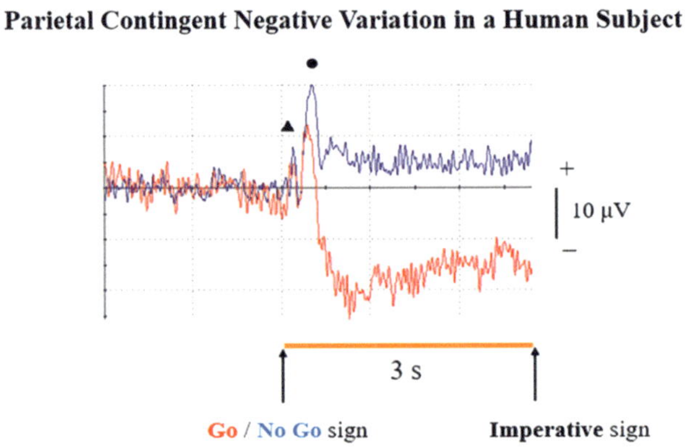

Figure 2.
Contingent negative variation (CNV) at the parietal area of the head in a human subject performing a warned Go/NoGo discrimination task with 3 s delay interval. The subject sitting on a chair, in response to the Go sign on the computer display, waits for 3 s (yellow horizontal line) and presses the button as quickly as possible when the imperative sign turns up on the display. In response to the NoGo sign, no response is required. Presentation of Go sign is followed by a sustained potential change of negative 15–20 µV (CNV). On the other hand, the No Go sign is followed by a static field potential change of around positive 5 µV. Go and NoGo signs are presented randomly and EEG is averaged 22 times for Go, and 10 times for NoGo signs in this recording. The average reaction time in this experiment is 367.2 ms. The filled triangle (▲) and filled circle (•) indicate P1 and P3 components of event-related potentials.

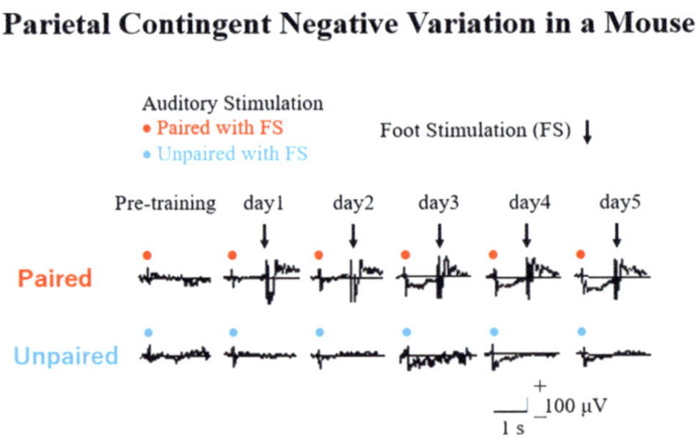

Figure 3.
Development of contingent negative variation (CNV) at the parietal region of a mouse in the training course to condition an auditory stimulation (•) to foot stimulation (↓). Another auditory stimulation of different frequency (•) is not followed by foot stimulation. Note that a negative shift of around 100 µV amplitude appears on day 3 of the training course.

Figure 4 shows a DC recording of EEG fluctuation at the frontal cortex in a behaving rat [13]. DC potential decreases when the rat is awake showing increased electromyogram (EMG) and decreased theta wave of EEG. The opposite profiles are observed when the arousal level is low. The amplitude of the DC potential fluctuation is around 2 mV.

These examples shown in **Figures 1–4** introduce the static and extremely low-frequency electric field on the head in humans and rodents recorded in various conditions and indicate that EEG amplitude in rats and mice is relatively large considering the small size of their brain (0.4–2 g) in comparison with the human brain weight

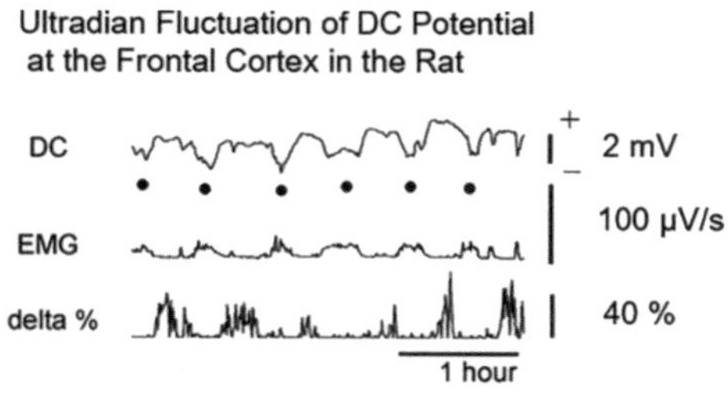

Figure 4.
Frontal DC potential recording in a rat showing an ultradian fluctuation. The DC potential becomes negative (•) when the amplitude of EMG increases and the amount of delta wave (delta %) decreases, indicating behavioral activation and an elevation of arousal. (modified from Shinba, 2009 [13]).

(1300–1400 g). Despite the size of the brain in rats and mice being about 1/1000 of that in humans, the generated electric field is the same or even larger than that in human. Taking the effects of skin impedance of the head in the human recording for consideration, the amplitude of the electric field is larger than expected from the size of the brain. This discrepancy between the brain size and the generated field potential amplitude may indicate that EEG recorded on the head represents the neural activity of discrete regions inside the brain, such as a certain part of the cortical layers. Localization of EEG generation can possibly lead to an estimation of electric field intensity larger than that assumed from the surface amplitude.

2.2 Localized electric field generation in the brain

The above discussion regarding the brain size suggests that the generation of surface EEG in the brain may be localized and can have an amplitude large enough to be functionally effective to influence the activity of the brain. The local electric field inside the brain is recorded with depth electrodes in clinical treatments in humans and in experimental research in animals to investigate electrical activity in relation to brain function [14]. It has been indicated that surface EEG mostly originates from local electric fields generated in the cortical layer structures producing dipole distribution with current sources and sinks [15]. Depending on the distribution of pyramidal neurons and interneurons in the layers of the cerebral cortex as well as on the dendritic configuration and inter-neuronal connections, various patterns of electric fields can be generated inside the cortex with current sources and sinks.

Figure 5 shows the electric field distribution in the auditory cortex of rats in response to tone stimulation [16]. In the anesthetized rat, tone stimulation initiates evoked responses in the auditory area of the temporal cortex with a peak latency of around 30 ms. The distribution of evoked potential is well localized in an area of the cortex (**Figure 5**, upper part). Surrounding the area showing the high evoked potential are the areas with less conspicuous potential. Under the areas with high surface evoked potential and with low surface evoked potential, different electric field profiles of auditory evoked potentials are observed in response to tone stimulation (**Figure 5**, lower part). Phase reversal of the fields indicates the site of potential generation inside the temporal cortex with a depth of 150–300 μm from the surface. When 150 μV potential is assumed to be generated between the layer interval of 150 μm, the electric field will be 1 V/m.

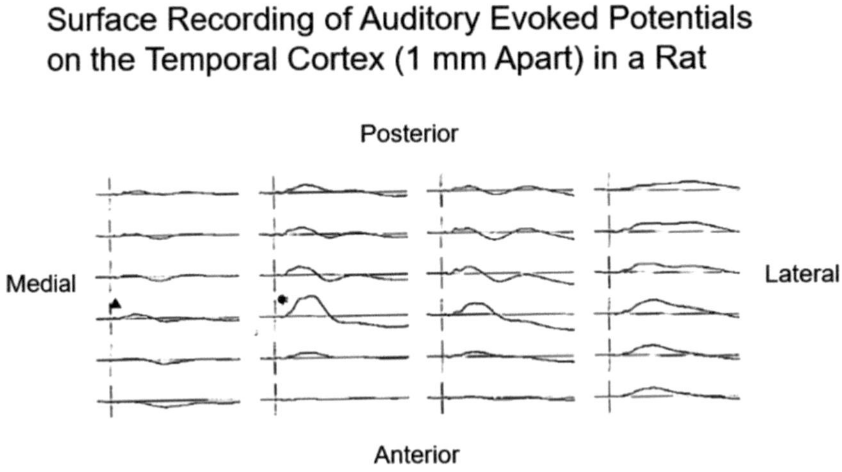

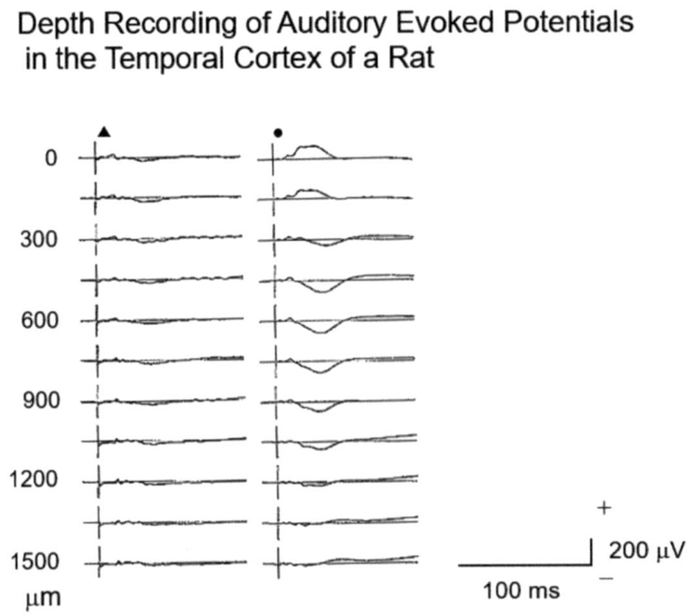

Figure 5.
Upper part: Topographical profiles of auditory evoked potential to tone stimulation (vertical bar) on the temporal cortex of an anesthetized rat. The recording sites are separated by 1 mm. Lower part: Depth recordings of auditory evoked responses to tone stimulation in the temporal cortex. The recording sites are separated by 150 μm. The left series of recordings are underneath the area showing a small surface potential (▲). The right series are under the area showing a large surface potential (•) (modified from Shinba et al., 1992 [16]).

The relations between the surface electric field, neuronal firing, and local electric field at the site of neuronal firing are presented in **Figure 6**. Clear relations are observed between the neuronal firings and electric field profiles with phase reversal at the site of neuronal firing. It is indicated that the temporal profile of endogenous electric field is closely related to the local neuronal activity. It is of interest to examine if the intensity of the electric field in the cortex around 1 V/m, intimately related

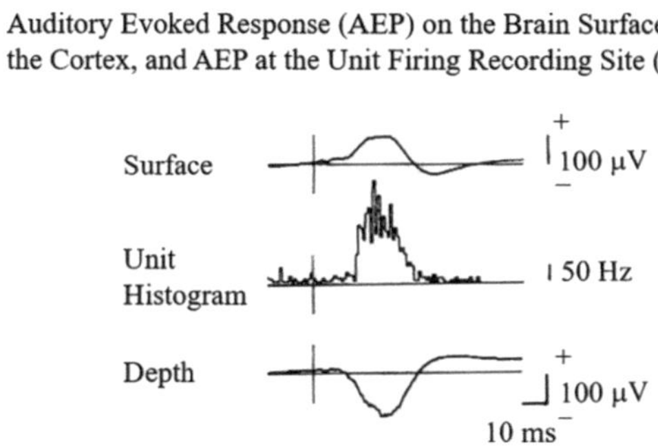

Figure 6.
Relation of surface auditory evoked potential (Surface), peri-stimulus time histogram of auditory evoked neuronal firing (Unit Histogram), and auditory evoked potential at the recording site of neuronal firing (Depth) (modified from Shinba et al., 1992 [16]).

to the evoked neuronal firing where the electric field is produced, is high enough to influence neuronal activities at other brain regions.

3. Effects of electric field on neural activity

3.1 Single-cell and slice preparation

At the single-cell level, previous research has reported that the application of an electric field of 0.1 mV/100 μm (1 V/m) influences the depolarization of cultured nerve cells [17]. The electric field of this intensity changes the transmembrane potential of a cell, leading to the facilitation or inhibition of the cell activities. In the studies using slice preparation of the nerve cells, electric field application of similar intensity is also known to entrain the electrical activities of the cells [18, 19]. The electric field of about 1 V/m in cultured cells and in slice preparation may serve to modify neuronal activities.

3.2 Electrical stimulation in the brain

In the *in vivo* experimental conditions, electrical stimulation in the brain is frequently employed to activate a discrete part of the brain to examine the effects of activation. It is known that electrical stimulation of a certain intensity can induce electric fields and depolarize neurons or neuronal fibers, altering the activities of target structures. The target activities include neuronal firing, release of transmitters and other substances, and movement of muscles and behavioral responses.

Figure 7 shows an example of an electrical stimulation method, which is frequently used in animal experiments. The distance between two electrode tips as well as the amount of current passed through the electrodes are different depending on the experimental settings. The parameters in **Figure 7** are employed in our previous study [10] to stimulate and depolarize the medial forebrain bundle in the rat brain to activate the reward system to study expectation. The effectiveness of the electrical stimulation with the employed intensity is proved by the appearance of the EEG waveform related to expectation. When the electrical resistivity of the brain tissue is

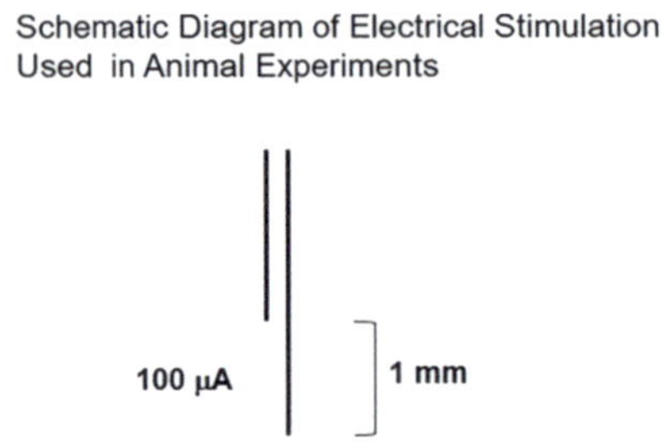

Figure 7.
A schematic diagram of electrical stimulation using paired electrodes with their tips separated by 1 mm. 100 μA current with 10 ms duration from one end of the electrode to the other end, 20 times with the inter-stimulus interval of 20 ms, can depolarize neuronal fibers in the rat brain [10]. When the resistivity is assumed to be 300 Ohm × cm [20], the resistance of the cubic tissue with one side being 1 mm is 3000 Ohm. The generated potential will be 300 mV between 1 mm distance. Then the electric field intensity is 300 V/m.

assumed as 300 Ohm × cm [20], the 100 μA current flow through the 1 mm × 1 mm plane with a distance of 1 mm will generate 300 V/m (**Figure** 7).

3.3 DC and AC stimulation to the brain

Clinically, electric field application on the body surface, such as direct current (DC) and alternating current (AC) stimulation, has been utilized for therapeutic purposes in humans. In the case of DC stimulation, 1–2 mA current is passed between two areas of the head for several minutes to modify the brain activity, although the underlying mechanisms are not fully clarified.

Figure 8 presents an example of DC stimulation (static electric field treatment) on EEG (endogenous electric field). This experimental setting can evaluate the effect of the externally applied static field to the brain on the waveform of endogenously generated event-related potential in the brain. The paradigm is the same as presented

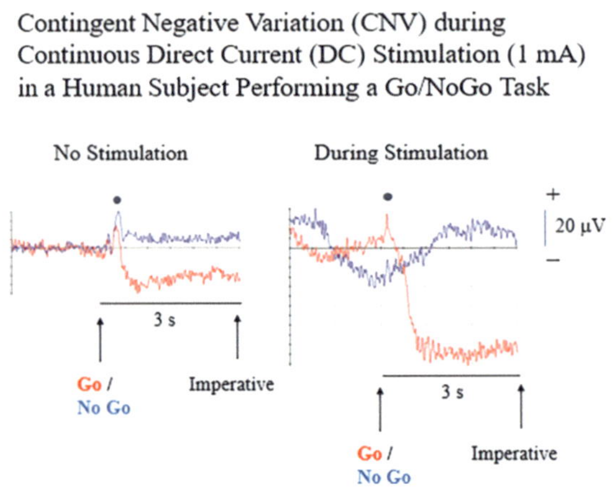

Figure 8.
Effects of DC electric field treatment between the frontal and occipital areas of the head on the contingent negative variation (CNV) in a human subject performing a Go/NoGo task. A filled circle (•) indicates the initial evoked potential. Continuous 1 mA DC treatment produces an increase in CNV amplitude. When the distance between the frontal and occipital electrodes is assumed to be 10 cm and the resistivity of the brain tissue is 300 Ohm × cm [20], then the electric field intensity between two sides of the 10 × 10 × 10 cm cube, estimating the brain size, will be 30 mV/10 cm (300 mV/m).

DOI: http://dx.doi.org/10.5772/intechopen.1006618

in the 2.1 subsection. The generated CNV waveform is found to be modulated under the background static electric field application. CNV amplitude increases during continuous DC stimulation under these experimental settings (**Figure 8**).

Similar effects are observed for the auditory evoked potential on the cortical surface during continuous static field application in an awake rat (**Figure 9**). The amplitude of the positive peak (upward triangle in **Figure 9**) decreases and that of the negative peak (reversed triangle in **Figure 9**) increases during DC application. It is interesting to note that the difference between the two peaks is not altered by the static field.

The applied electric fields used in these two observations are lower than that conventionally employed as electrical stimulation introduced above in **Figure 7**. It may be suggested that the applied electric fields do not depolarize neurons to transmit the signals through synapses to other brain regions to produce modification of endogenous waves, but would alter the brain activity through ephaptic coupling. Further research is necessary to find the mechanisms underlying the observations.

3.4 Exogenously treated noninvasive electric field and EEG

Exogenous electric field surrounding the body, inducing internal electric field also has effects on internal activities. The intensity of the internally generated electric field is low due to the cage effect of the body. **Figure 10** shows that the 1 min exogenously applied 50-Hz 30-kV electric field between the electrodes, one above the head and the other underneath the feet, induces the internal electric field of less than 1 V/m inside the body, and concomitant changes in the EEG power are generated [21, 22]. An increase of the theta (4–7 Hz) EEG power suggests an increased relaxation of the subject. The mechanism underlying this finding is not clear but may involve a direct effect of exogenously-generated internal electric field on the brain endogenous electric field or an indirect effect through other structures. The exogenous and endogenous interaction would be an interesting target of research to understand the electric profiles of the human nervous system.

Effects on the skin can be the candidate for the mechanism of action [23]. Extremely low-frequency electric field causes vibration of skin and hair causing dilatation of blood vessels and increase of blood flow. The information of cutaneous

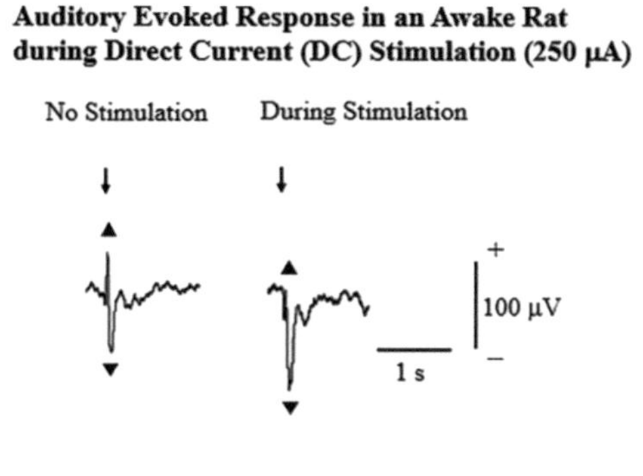

Figure 9.
Effects of DC electric field treatment between the frontal and the cerebellum on the auditory evoked potential at the parietal region in an awake rat (↓ tone stimulation). Continuous 250 μA DC current produces alteration of the waveform of the auditory evoked potential (▲ positive peak, ▼ negative peak). During DC stimulation, the negative peak of the waveform increases its amplitude, and the positive peak decreases it. When the distance between the frontal and cerebellum electrodes is assumed as 1 cm and the resistivity of the brain tissue as 300 Ohm × cm [20], then the electric field intensity between two sides of the 1 × 1 × 1 cm cube will be 75 mV/cm (7.5 V/m).

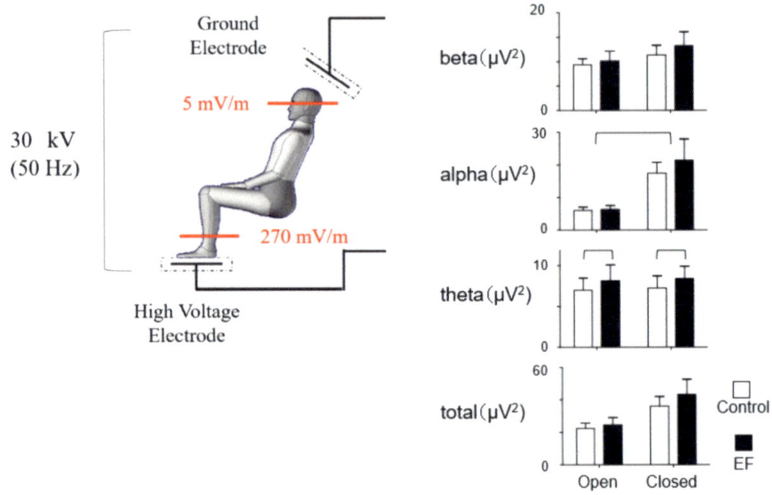

Figure 10.
Externally applied extremely low-frequency electric field treatment for 1 min (30 kV, 50 Hz) produces an internal electric field of 5–270 mV/m in the body when calculated by dosimetry using a phantom, which is associated with EEG theta power increment in the human subjects during both eyes open and closed states. Alpha power is significantly greater during the eyes-closed state. The data connected by a line show statistically significant difference (modified from Shinba et al. [21, 22]).

changes is sent to the brain by afferent nerves to produce modification of psychophysiological state. The skin itself generates an electric field as described below, which may be affected by the exogenous electric field. Internally generated electric field may also alter hormonal activities, leading to alteration of arousal reflected in EEG changes. Harakawa and colleagues [24] have reported the effects of electric fields on endocrinological system related to stress. Various structures in the body can interact with exogenously generated electric fields, producing the changes in endogenous electric fields.

Interaction with the naturally and artificially generated exogenous electric field is also necessary to be considered to assess the effects on the endogenous electric field. The frequency of the electric field outside the body is not only in the range of static and extremely low state but includes the higher ranges, such as mega-, tera-, and up to exa-hertz. In the higher frequency ranges, magnetic fields are concomitantly generated together with electric fields, producing complex effects on human and animal biological functions [1–4]. It is necessary to examine the magnetic field effects when considering the exogenously-generated internal electric field.

4. Electric fields in the muscles and skin

4.1 Muscle potential

The membrane potential of muscle fibers is altered to initiate muscle contraction. When the muscle fibers are depolarized, their transmembrane potential, which is more than 50 mV, is reduced by the influx of calcium ions. This triggers the structural changes of muscle fibers including myosin and actin causing muscle contraction. Electrical energy stored as transmembrane potential is transformed into mechanical energy. At the same time, electric fields are generated in the muscle structure.

The profiles of electric fields are different depending on the muscles, including the skeletal, heart, and gastrointestinal muscles, but all produce mV-order field potential when measured at the muscle surface or on the body surface [25].

The heart produces an electric field which is recorded as an ECG on the chest. The R wave corresponds to the initiation of ventricular muscle depolarization from the inner to the outer layer of the muscle wall. The amplitude of the R wave is around 1 mV and the electric field will be 50 mV/m when the wall is 2-cm thick [26].

Skeletal muscle also produces a surface potential of around 1 mV. When the potential is assumed to be generated between the two sides of the muscle with the distance assumed to be 1 cm, the electric field will be around 100 mV/m.

The stomach muscle generates electric field which is recorded as an electrogastrogram on the abdominal skin surface [25]. The fields are related to the movement of the stomach. The peristaltic movement frequency viewed by electrogastrogram is around 3 times per min and the field amplitude increases after the meal. The recorded potential is 10–30 mV and corresponds to 2–6 V/m when the stomach wall is 5-mm thick.

4.2 Skin potential

Skin potential is an important component of electric field endogenously generated in the body. Its amplitude is around 30 mV when recorded on the scalp [27], and more than 10 mV when recorded between the finger and the arm [28]. It also fluctuates depending on the behavioral state. When the thickness of the skin is assumed to be 2 mm, the electric field intensity is 5–15 V/m. The function of cutaneous electric field is related to cell migration and wound healing [29]. The generation mechanisms of skin potential may include the membrane potential at eccrine sweat glands but are not thoroughly clarified.

Effects of exogenously applied electric fields on various internal organs may involve the changes in skin potential. Skin potential would be important in considering the role of skin as the cage and capacitance for an exogeneous electric field. By passing some current through the skin, the induced current on the inner side of the skin is reduced. As the capacitance, the dielectric polarization also reduces the electric field generated internally. Skin is not a highly conductive material nor a full insulator. Conductivity also changes depending on the physical conditions such as sweating or humidity. Modifications of skin conditions may control the effects of the exogenous electric field on the endogenous electric field.

4.3 Electric field modulation by physical changes

Modulation of endogenous electric fields can also occur in response to various physical changes in the body, including gastrointestinal peristatic movements and pressure changes caused by heartbeats. Gastrointestinal movements are continuous and should produce complex changes in electric fields. The functional significance of these changes has not been fully studied, and the electric field rhythm can be a marker for integrated gastrointestinal activity.

As for the cardiac movement, each heartbeat pumps out about 70 mL of blood from the ventricle to the peripheral structures, causing pressure in the peripheral tissue. Because the impedance of blood is low in comparison with other tissues, rhythmic changes of arterial blood volume can influence the endogenous electric field intensity.

Movements of skeletal muscle and skin accompanying body movement also change the endogenous electric field structure. Future studies are necessary to examine the functional significance of electric field changes related to muscle and skin movements.

5. A viewpoint on ephaptic physiology

The data presented above indicate that in the body including the brain of humans and animals, static and extremely low-frequency endogenous electric fields are generated with the intensity high enough to influence neuronal activities, enabling ephaptic interaction. On the human head surface, the amplitude of electric field is at μV-order and electric field appears to have little functional significance. However, the amplitude of EEG on the skull surface is not dependent on the brain size, and when viewed inside the brain of rodents, the potential is localized in particular areas, including the upper layer of the cortex, showing around 1 mV difference between the distance of 1 mm. The electric field of about 1 V/m is generated locally inside the brain. An electric field intensity of 1 V/m is not too weak to be functional. Single-cell and slice studies indicate that electric field of this intensity can act on the excitability of neurons. Entrainment of activity by electric field of around 1 V/m is also observed in the neural network system [7, 8].

The presence of ephaptic interaction is also supported by the data showing an effect of externally applied static electric field on endogenously generated EEG. Enhancement of EEG power by externally applied extremely low-frequency electric field can also be viewed as ephaptic coupling [21, 22]. The present chapter has mostly discussed the functional perspectives of electric fields in the brain, but also has referred to the electric fields in the muscle and skin structures. The intensity of the electric field in the muscle and skin is comparable to that in the brain and may have the potential to serve as functional information.

The data reviewed above in this chapter support the hypothesis that electric fields could operate in the body of living organisms including human beings to functionally communicate with diversely situated structures and to integrate the activities. **Figure 11** can be the example of ephaptic interaction in the brain. It is shown that the EEG with the frequency of delta to theta range (0.5–7 Hz, AC in **Figure 11**) changes its amplitude in parallel with the slower EEG wave with the frequency of about 0.02 Hz (DC in **Figure 11**). Two types of electric fields with different frequencies are

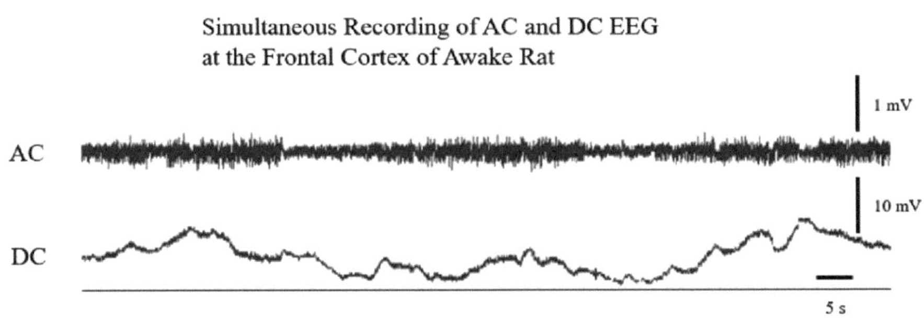

Figure 11.
Both AC and DC EEG are simultaneously recorded from the same electrode on the frontal cortex of an awake rat. Note that a large AC fluctuation is accompanied by positive shift in static electric field.

interrelated, suggesting the possibility of ephaptic coupling, although future studies are necessary to validate this interpretation.

It has been discussed that endogenously generated electric field is not an epiphenomenon. The endogenous electric field is around 1 V/m at various sites of the body in humans and rodents, and the intensity can be large enough to play physiological roles. Nerve conduction and synaptic transmission may be too slow for crediting the organic integrity or consciousness. The coherent activity of brain and peripheral structures could be possible through ephaptic interaction. In addition to the conventional mode of information based on the neuronal activation and synaptic transmission, non-synaptic transmission of information through electric fields may play a significant role. Ephaptic physiology will be an important area of research to understand the nature of integrative processes.

Acknowledgements

The author thanks Shinji Harakawa of Obihiro University of Agricultural and Veterinary Medicine, Takashi Nedachi of Hakuju Institute for Health Science Co., Ltd., Go Mugishima of Fukuoka Prefectural University, Yoshiki Matsuda of Tokyo Institute of Medical Sciences, Yujiro Shinba of Chiba Emergency and Psychiatric Medical Center, and Shuntaro Shinba of Shimada General Medical Center for their important comments and suggestions on this chapter.

Conflict of interest

The author declares no conflict of interest.

Author details

Toshikazu Shinba
Department of Psychiatry, Shizuoka Saiseikai General Hospital, Shizuoka, Japan

*Address all correspondence to: t156591@siz.saiseikai.or.jp

References

[1] International Commission on Non-Ionizing Radiation Protection. Guidelines for limiting exposure to time-varying electric and magnetic fields (1 Hz to 100 kHz). Health Physics. 2010;**99**:818-836

[2] International Commission on Non-Ionizing Radiation Protection. Guidelines for limiting exposure to time-varying electric, magnetic, and electromagnetic fields (up to 300 GHz). International commission on non-ionizing radiation protection. Health Physics. 1998;**74**:494-522

[3] World Health Organization. Extremely Low Frequency Fields Environmental Health Criteria Monograph No. 238. Geneva, Switzerland: World Health Organization; 2007. Available from: https://www.who.int/publications/i/item/9789241572385

[4] Cook MR, Graham C, Cohen HD, Gerkovich MM. A replication study of human exposure to 60-Hz fields: Effects on neurobehavioral measures. Bioelectromagnetics. 1992;**13**:261-285

[5] Niedermeyer E. Eplileptic Seisure Disorders. In: Niedermeyer E, Lopes Da Sylva F, editors. Electroencephalography: Basic Principles. Clinical Applications, and Related Fields. 5th ed. Philadelphia, PA: Lippincott Williams and Wilkins; 2005. pp. 505-619

[6] Martin AR, Brown DA, Diamond ME, Cattaneo A, De-Miguel FF, Nicholls J. From Neuron to Brain. 6th ed. Sunderland, MA, USA: Sinauer Associates Inc.; 2021

[7] Anastassiou CA, Koch C. Ephaptic coupling to endogenous electric field activity: Why bother? Current Opinion in Neurobiology. 2015;**31**:95-103. DOI: 10.1016/j.conb.2014.09.002

[8] Fröhlich F. Endogenous and exogenous electric fields as modifiers of brain activity: Rational design of noninvasive brain stimulation with transcranial alternating current stimulation. Dialogues in Clinical Neuroscience. 2014;**2014**(16):93-102. DOI: 10.31887/DCNS.2014.16.1/ffroehlich. 16: 1

[9] Caspers H, Schulze H. Changes in the direct cortical potential during the natural sleeping-waking period in the freely-moving animal. Pflügers Archiv für die gesamte Physiologie des Menschen und der Tiere. 1959;**270**:103-120

[10] Matsuda Y, Ozawa N, Shinozaki T, Tatebayashi Y, Honda M, Shinba T. Physiological paradigm for assessing reward prediction and extinction using cortical direct current potential responses in rats. Scientific Reports. 2024;**14**:10422. DOI: 10.1038/s41598-024-59833-7

[11] Fuster JA. The Prefrontal Cortex. 5th ed. Cambridge, MA, USA: Academic Press; 2015

[12] Shinba T, Sumi M, Iwanami A, Ozawa N, Yamamoto K. Increased neuronal firing in the rat auditory cortex associated with preparatory set. Brain Research Bulletin. 1995;**37**:199-204

[13] Shinba T. 24-h profiles of direct current brain potential fluctuation in rats. Neuroscience Letters. 2009;**465**:104-107

[14] Shinba T, Andow Y, Shinozaki T, Ozawa N, Yamamoto K. Event-related potentials in the dorsal hippocampus of

rats during an auditory discrimination paradigm. Electroencephalography and Clinical Neurophysiology. 1996;**100**:563-568

[15] Mitzdorf U. Current source-density method and application in cat cerebral cortex: Investigation of evoked potentials and EEG phenomena. Physiological Reviews. 1985;**65**:37-100. DOI: 10.1152/physrev.1985.65.1.37

[16] Shinba T, Ando Y, Ozawa N, Yamamoto K. Auditory-evoked response of the cortex after yohimbine administration: Phase advance effect of central noradrenergic activation. Brain Research Bulletin. 1992;**28**:463-471

[17] Terzuolo CA, Bullock TH. Measurement of imposed voltage gradient adequatae to modulate neuronal firing. Proceedings of National Academy of Sciences USA. 1956;**42**:687-694. DOI: 10.1073/pnas.42.9.687

[18] Bikson M, Inoue M, Akiyama H, Deans JK, Fox JE, Miyakawa H, et al. Effects of uniform extracellular DC electric fields on excitability in rat hippocampal slices in vitro. Journal of Physiology. 2004;**557**(Pt. 1):175-190. DOI: 10.1113/jphysiol.2003.055772

[19] Asamoah B, Khatoun A, Mc LM. Frequency-specific modulation of slow-wave neural oscillations via weak Exogeneous extracellular fields reveals a resonance pattern. Journal of Neuroscience. 2022;**42**:6221-6231. DOI: 10.1523/JNEUROSCI.0177-22.2022

[20] Geddes LA, Baker LE. The specific resistance of biological materials. Medical and Biological Engineering. 1967;**5**:271-293

[21] Shinba T, Nedachi N, Harakawa S. Extremely low-frequency electric field exposure increases theta power of EEG in both eyes-open and eyes-closed resting conditions in healthy male subjects. IEEJ Transactions on Electrical and Electronic Engineering. 2021;**16**:592-599. Pub Date: 2021-02-24. DOI: 10.1002/tee.23334M

[22] Shinba T, Nedachi N, Harakawa S. Alterations in heart rate variability and electroencephalogram during 20-minute extremely low frequency electric field treatment in healthy men during the eyes-open condition. IEEJ Transactions on Electrical and Electronic Engineering. 2022, 2022;**18**:38-44. DOI: 10.1002/tee.23695

[23] Shimizu K, Endo H, Matsumoto G. Fundamental study on measurement of ELF electric field at biological body surfaces. IEEE Transactions on Instrumentation and Measurement. 1989;**38**:779-784

[24] Harakawa S, Nedachi T, Shinba T, Suzuki H. Stress-reducing effect of a 50 Hz electric field in mice after repeated immobilizations, electric field shields, and polarization of the electrodes. Biology (Basel). 2022;**11**:323. DOI: 10.3390/biology11020323

[25] Geldof H, Van Der Schee EJ, Van Blankenstein M, Grahsuis JL. Electrogastrographic study of gastric myoelectrical activity in patients with unexplained nausea and vomiting. Gut. 1986;**27**:799-808

[26] Yalin K, Aksu T, Ikitimur B, Onder SE, Soysal AU, Ozturk S, et al. Intracardiac electrogram characteristics of intramural outflow tract ventricular arrhythmias. Journal of Interventional Cardiac Electrophysiology. 2023;**66**:621-627. DOI: 10.1007/s10840-022-01374-y

[27] Picton YW, Hillyard SA. Cephalic skin potentials in electroencephalography cephalic skin. Electroencephalography and Clinical Neurophysiology.

1972;**33**:419-424. DOI: 10.1016/0013-4694(72)90122-8

[28] Li Y, He J, Fu C, Jiang K, Cao J, Wei B, et al. Children's pain identification based on skin potential signal. Sensors (Basel). 2023;**23**:6815. DOI: 10.3390/s23156815

[29] Funk RHW. Endogenous electric fields as guiding cue for cell migration. Frontiers in Physiology. 2015;**6**:143. DOI: 10.3389/fphys.2015.00143

Chapter 7

Modeling and Optimal Design Strategies for Wire-Plate ESP

Mohamed Badran and Abdallah Alkhaiyat

Abstract

This chapter focuses specifically on wire-plate-type electrostatic precipitators (ESPs), detailing their fundamental operation principles and the role of electrostatic forces in particle collection. It also outlines the key mathematical models used to simulate ESPs, providing insights into the interactions between electric fields, airflow, and particle dynamics. Furthermore, the chapter investigates the impact of geometric variations—such as electrode spacing, plate length, and channel height—on wire-plate ESP efficacy by utilizing finite element analysis. The findings indicate that optimizing these geometric parameters can substantially enhance collection efficiency, reduce operational costs, and mitigate potential particle re-entrainment. In an industrial context, several key parameters must be considered to maximize the efficiency and cost-effectiveness of the ESP. Factors such as power consumption, inlet flow velocity, specific collection area, plate-to-plate distance, and corona power ratio influence the overall performance of the system. The chapter concludes with practical design and optimization guidelines, demonstrating that an optimal balance between high collection efficiency and low energy usage can be achieved, thereby enhancing the environmental and economic performance of wire-plate ESPs.

Keywords: electrostatic precipitators, wire-plate, particulate emissions control, electrostatic forces, particle collection, mathematical modeling, simulation, finite element analysis, design optimization, energy consumption

1. Introduction

Particulate emission control technologies are essential for preserving environmental quality and public health [1–3]. These technologies are engineered to capture particulate matter (PM) before they are released into the atmosphere from industrial processes. Electrostatic precipitators (ESPs) are air pollution control devices that utilize electrostatic forces to remove PM from gas streams [4]. They are widely used for cleaning exhaust gases in industrial settings, primarily due to their efficiency in capturing a broad spectrum of particles ranging from 0.01 to 10 μm with high efficiency [5–7]. This capability has been particularly significant in recent years, as environmental regulations on atmospheric particle emissions have become increasingly stringent [8, 9]. By minimizing PM emissions, ESPs contribute to cleaner air and the mitigation of respiratory diseases [4].

IntechOpen

ESPs remove particulate matter from flue gas by using an electrical field that ionizes the particles, followed by their transport and collection [10]. Although the concept behind ESPs is well-established [11], their application remains crucial in modern pollution control. In this section, different classifications of ESPs are covered, followed by an introduction to the working principles of ESPs.

1.1 ESP classification

Electrostatic precipitators can be categorized based on different criteria including the number of stages, or the collection plate configuration [2]. This section presents an overview of ESP design, and how different configurations are classified.

1.1.1 Classification based on the collecting electrode design

The arrangement of collecting electrodes can differ significantly between various electrostatic precipitators. The wire-plate type configuration, depicted in **Figure 1** [12], features parallel collection plates that surround the emitting electrodes. These electrodes, which are also called discharge electrodes, are responsible for generating the corona discharge. In this configuration, the emitting wire radius is denoted as r_W, and the spacing between adjacent wires is denoted as S. The plate-to-plate distance and the width of the collection plate are given by d and W_p, respectively.

The wire-plate design is well-suited for large volumes of flue gas, as it allows for multiple parallel flow lanes, with each lane passing the discharge wires in sequence as the gas flows through the ESP [13, 14].

Transverse plate ESPs, such as the one illustrated in **Figure 2a** [12], are another type that uses parallel collecting plates. However, in transverse plate ESPs, these plates are arranged perpendicular to the airflow and are evenly spaced at a distance d. The primary advantage of this design is that the transverse arrangement enhances the momentum of ions significantly, enabling them to overcome the electric field more effectively and thus improving particle transport [15], as the charged particles are more effectively directed toward the collection plates, and subsequently increasing the overall collection efficiency of the ESP. L_p here is the length of the collection plate.

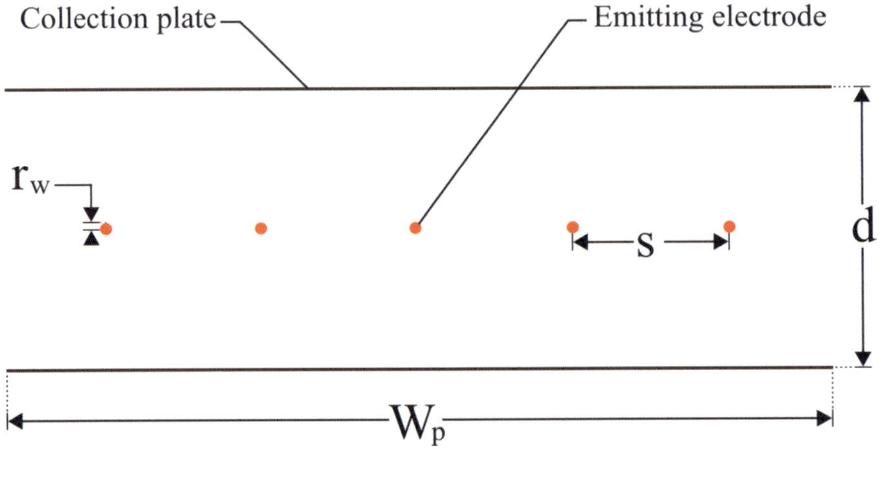

Figure 1.
Schematic of a single-channel wire-plate ESP [12].

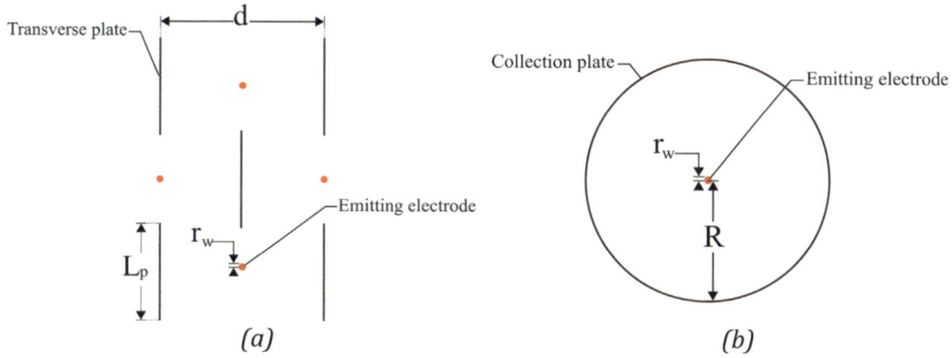

Figure 2.
ESPs with different electrode configurations: (a) transverse plate type; (b) wire-cylinder type [12].

Another type is the wire-cylinder type ESP, also known as the tubular ESP. The main dimensions, shown in **Figure 2b** [12], are the cylindrical collection plate radius, *R*, and the emitting electrode length. A cylindrical tube serves as the collection plate in this design, with emitting electrode(s) placed at its center. The flue gas enters from one end of the vertical cylinder and flows along the direction of the cylindrical tube, perpendicular to the cross-section depicted in **Figure 2b** [12]. The collected particles then settle at the bottom of the tube. The performance of wire-cylinder ESPs depends on factors like the radius, *R*, of the collection plate, the length of the emitting electrodes, and the gas flow rate. This configuration is effective in removing PM in limited spaces where high dust loads are encountered.

1.1.2 Classification based on the number of stages

A stage in the context of ESPs refers to a distinct phase in the particle removal process. In a single-stage ESP, both the ionization (charging) of particles and their collection occur within the same section [10]. Conversely, a multi-stage ESP separates these processes into distinct stages, with one section dedicated to charging the particles and another dedicated to their collection [12]. The primary advantage of single-stage ESPs, other than being simple, is their reduced susceptibility to particle re-entrainment [16, 17].

On the other hand, multi-stage ESPs typically feature larger collecting electrode surface areas due to the reduced spacing between the electrodes and the allocation of a dedicated section for particle collection. Another advantage of this design is its resilience to variable inlet flow velocities, as having a dedicated charging section can better accommodate fluctuations in the inlet gas flow rates. This design still maintains a compact footprint like single-stage ESPs. Separating the charging and collection stages offers additional benefits, including improved precipitation efficiency [18, 19].

2. Fundamental principles of wire-plate ESPs

ESPs utilize electrostatic forces to remove particles, setting them apart from conventional techniques such as mechanical filtration. In this section, we will explore the key components of ESPs and provide an overview of their operating principles.

2.1 Major components of ESPs

The configuration of an ESP comprises multiple discharge electrodes, typically thin wires, alongside collecting electrodes arranged as parallel plates in wire-plate ESP designs. These components, illustrated in **Figure 3**, facilitate the removal of particulate matter from gas streams.

Upon entering the ESP through an inlet duct designed to minimize turbulence, the gas flows past the wire electrodes, which generate a strong electric field. This section is often referred to as the ionization stage. During ionization, particles within the gas become charged, enabling their migration toward the collection plates. As these charged particles accumulate on the collection plates, a rapping mechanism periodically strikes the plates to dislodge the collected particles, which then fall into a hopper for collection.

The collection plates are often segmented, allowing for independent rapping of each section. This periodic cleaning is key to maintaining precipitation efficiency. Finally, the cleaned gas is expelled into the atmosphere through the outlet duct.

2.2 Operating principles of ESPs

The operation of an ESP is fundamentally based on a series of processes that facilitate the collection of particles from a gas stream. These processes include corona discharge, ionization of particles, and their subsequent transport toward collecting electrodes. The entire system is powered by a High-Voltage Direct Current (HVDC) power supply that powers both the charging and collection stages. The polarity of the plates in the collection stage is opposite to that of the wires in the charging stage, as depicted in **Figure 4**. The HVDC system converts standard industrial alternating current (AC) voltage into a pulsating direct current (DC) voltage, typically ranging from 20 to 100 kV.

Upon entering the ESP, the flue gas flows through the ionization stage, where the HVDC power supplies discharge wires. This process continues until the electric field

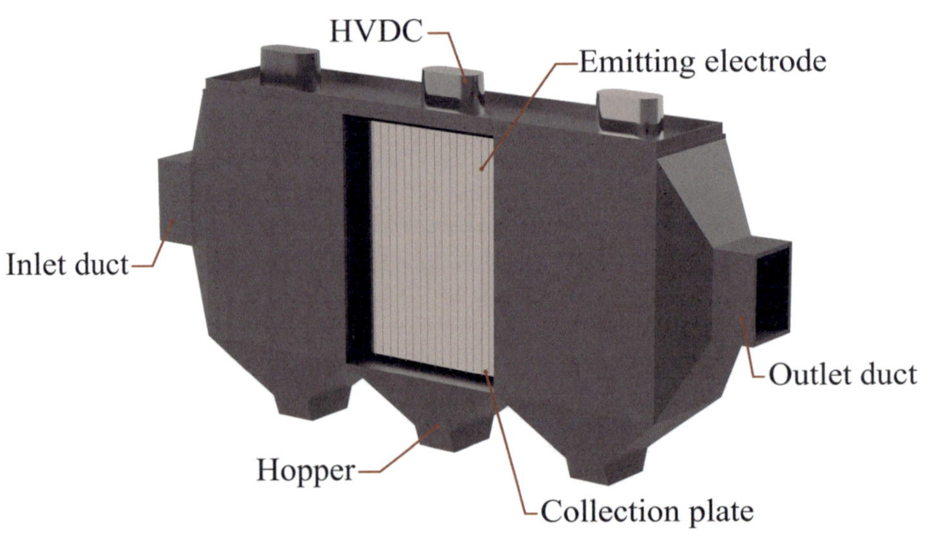

Figure 3.
Major components of a wire-plate ESP.

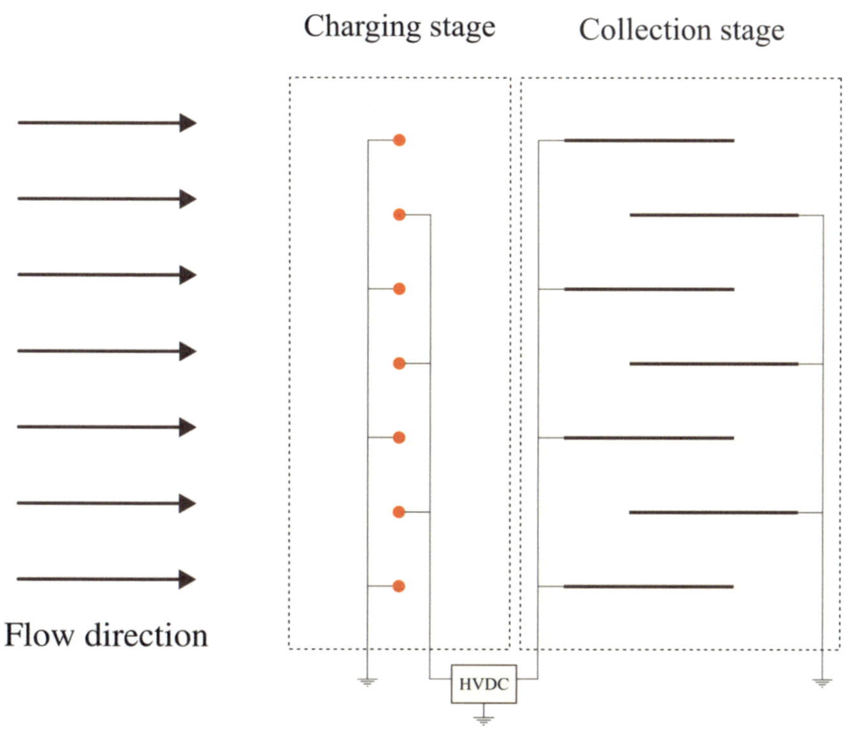

Figure 4.
Two separate charging and collection stages in a two-stage ESP [12].

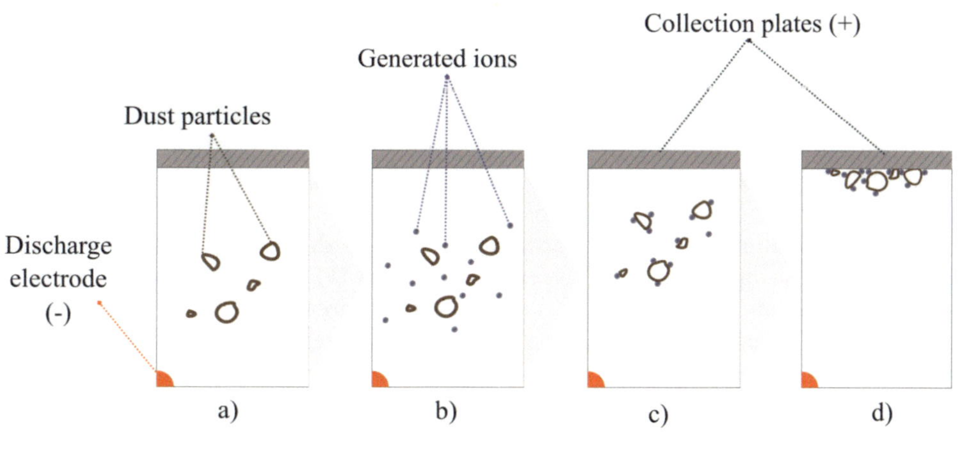

Figure 5.
Particle collection steps in an ESP. (a) Flue gas; (b) Corona discharge; (c) Particle ionization; (d) Particle transport.

reaches a specific value, known as the electric field threshold value, as shown in **Figure 5a**. At this point, corona discharge onsets when the electrons are stripped from the flue gas particles as shown in **Figure 5b**. This ionization process creates a cloud of ions in the surrounding gas, which subsequently attach to the flow particles as can be seen in **Figure 5c**. Notably, this interaction includes a diffusion charging region for particles smaller than 0.1 μm, while larger particles exceeding 1 μm fall into the field charging region. Practically, particles experience both types simultaneously [20].

Once charged, the particles migrate toward the collecting plates, which possess an opposite polarity, as illustrated in **Figure 5d**. The electrostatic forces generated by the electric field drive this migration, effectively pulling the charged particles toward the plates. The design of the plates maximizes the surface area for particle accumulation, thereby enhancing the overall efficiency of the ESP.

Over time, the accumulation of particles on the collection plates can reduce the efficiency of the ESP. The collected particles are then dislodged by activating a rapping mechanism, allowing them to fall into the hopper below. Alternative methods for dislodging particles include scraping the collection plates or washing them off with water. This cleaning process is essential for maintaining optimal performance, as accumulated particles can be re-entrained after collection into the gas flow if not properly managed, which would negatively impact the collection efficiency of the ESP [20].

3. Mathematical modeling and its application on COMSOL multiphysics

Mathematical modeling serves as a crucial framework for analyzing and optimizing the performance of ESPs. Modeling ESPs involves several approaches ranging from empirical to computational models. This section explores the various modeling approaches employed in the study of ESPs, followed by a case study detailing how Finite Element Analysis (FEA) can be implemented using COMSOL Multiphysics for a wire-plate ESP.

3.1 Overview of modeling approaches

The first approach is to use empirical models based on correlating experimental data to practical observations. These models often simplify the complex interactions within the ESP into empirical relationships. However, this approach generally lacks the accuracy found in more detailed models.

Computational models utilize mathematical relationships to accurately simulate ESPs by solving coupled partial differential equations (PDEs) numerically. These models typically involve solving for the electrostatic forces, charging of particles, and airflow. This method requires defining the problem by setting up boundary conditions, such as the inlet and outlet conditions for airflow, as well as initial conditions for electrode potentials and particle distribution.

3.2 Electrostatic forces and particle charging

This section gives an overview of the governing equations for modeling ESPs, where key equations for the electrostatic forces and particle charging are presented. The corona discharge phenomenon results from the ionization of gas molecules that surround the discharge electrodes. To model the electrical field inside of an ESP, Poisson's equation [21, 22] can be used:

$$\nabla \cdot \vec{E} = \rho/\varepsilon_0 \tag{1}$$

Where $\vec{E}$ is the electric field intensity (V m^{-1}), ρ is the ionic space charge density (C/m^3), and ε_0 is the vacuum permittivity (F m^{-1}). The electric field is related to the electric potential, V (V), with the following relation:

$$\vec{E} = -\nabla \cdot V \tag{2}$$

Electric current density is a combination of three mechanisms, namely: convection, conduction, and diffusion, and is represented by:

$$\vec{J} = \mu_e \, \rho \, \vec{E} + \rho\vec{U} - D \cdot \nabla\rho \tag{3}$$

Where $\vec{J}$ is the current density (A m^{-2}), μ_e is the ionic mobility (m^2 V^{-1} s^{-1}), $\vec{U}$ is the airflow velocity, and D is the diffusion coefficient (m^2 s^{-1}). Conduction term will be neglected in this study [23], leading to the following relation:

$$\vec{J} = \mu_e \, \rho \, \vec{E} - D \cdot \nabla\rho \tag{4}$$

As DC voltage is applied to the electrodes of an ESP, charge conservation is satisfied under steady-state conditions and current continuity is therefore:

$$\nabla \cdot \vec{J} = 0 \tag{5}$$

Peek's law is an empirical relationship that is used to obtain the threshold electric field as follows [24, 25]:

$$\vec{E_i} = f \times 3.1 \times 106\left(\delta + 0.0308 \cdot \sqrt{\frac{\delta}{r_w}}\right) \tag{6}$$

Where $\vec{E_i}$ is the threshold electric field value (V m^{-1}), f is the wire roughness factor, δ is the density of the air at 1 (atm) and 298 (K) [25, 26], and r_w is the emitting wire radius (m).

Equations (1) and (2) can be reduced to:

$$\nabla^2 \cdot V = -\rho/\varepsilon_\circ \tag{7}$$

Starting with (4) and (5), we can deduce the following:

$$\nabla \cdot \vec{J} = \nabla \cdot \left(\mu_e \, \rho\vec{E}\right) = 0 \tag{8}$$

$$\mu_e \cdot \left(\vec{E}\nabla \cdot \rho + \rho\nabla \cdot \vec{E}\right) = 0$$

$$\mu_e \cdot \left(\vec{E}\nabla \cdot \rho\right) = -\mu_e \cdot \left(\rho\nabla \cdot \vec{E}\right)$$

$$\vec{E}\nabla \cdot \rho = -\rho\nabla \cdot \vec{E} \tag{9}$$

By using (1) and (9):

$$\vec{E}\nabla \cdot \rho = -\rho^2/\varepsilon_\circ \tag{10}$$

Using (2) and (10) finally yields:

$$0 = \nabla(V) \cdot \nabla(\rho) - \rho^2/\varepsilon_\circ \tag{11}$$

3.3 Airflow patterns

Crucially, the flow inside of ESPs is turbulent. The turbulent flow can be described using Reynolds averaged Navier-Stokes equations for incompressible gas using the following formula [22]:

$$\rho_g\left(\vec{U} \cdot \nabla\right)\vec{U} = -\nabla p + \mu_g \nabla^2 \vec{U} + F \tag{12}$$

Continuity of gas flow:

$$\nabla\vec{U} = 0 \tag{13}$$

Where ρ_g is the density of the flue gas (kg m^{-3}), p is the fluid pressure (N m^2), μ_g is the dynamic viscosity the flue gas (kg m^{-1} s^{-1}), and F represents the external forces applied to the fluid.

Considering that the pressure drop is small across ESPs, the air is assumed to be a steady, incompressible Newtonian fluid [27]. Meaning that Eqs. (12) and (13) can be solved by the conservation of mass equation:

$$\frac{\partial \rho_g}{\partial t} + \nabla \cdot \left(\rho \vec{U}\right) = 0 \tag{14}$$

Conservation of momentum is given by [28]:

$$\vec{U} \cdot \nabla\vec{U} = -\frac{\nabla P}{\rho_g} + \left(\mu_g + \mu_t\right)\nabla^2\vec{U} + \rho_i\vec{E} \tag{15}$$

Where ρ_i is the ion charge density (C m^{-3}), and μ_t is the turbulent dynamic. viscosity (kg m^{-1} s^{-1}).

3.4 Particle dynamic

Particles within the flue gas are influenced by four forces, starting with the electrostatic force [29, 30] $\vec{F_e}$, and the gravity force [31], given by:

$$\vec{F_g} = -\rho_{par} \times V_{par} \times \vec{g} \tag{16}$$

where ρ_{par} is the density of the particle (kg m^{-3}), V_{par} is the volume of the particle (m^3), and $\vec{g}$ is the acceleration due to gravity (m s^{-2}). Archimedes force is given by [32]:

$$\vec{F_a} = -\rho_g \times V_{par} \times \vec{g} \tag{17}$$

The drag force can be modeled by the empirical Khan-Richardson model [33] given that the particles' impact on the flow field is negligible.

$$\overrightarrow{F_d} = \pi a^2 \rho_g \left(\vec{U} - \vec{U}_{par}\right)^2 \left[1.84\ Re_p^{-0.31} + 0.293\ Re_p^{0.06}\right]^{3.45} \tag{18}$$

where a is the particle radius (m), $\vec{U}_{par}$ is the particle velocity (m s^{-1}), and Reynolds number is defined by:

$$Re_p = \frac{\left(\vec{U} - \vec{U}_{par}\right) 2a\rho_g}{\mu_g} \tag{19}$$

3.5 Particle collection efficiency

Particle collection performance of the ESP can be evaluated by the collection efficiency, which is the ratio of the mass of the collected particles, m_{out} (kg), to the mass of the input particles, m_{in} (kg) [34]:

$$\eta = 1 - \frac{m_{out}}{m_{in}} \tag{20}$$

It can also be expressed as the ratio of the concentration of the collected dust, C_{out} (kg m^{-3}), to the concentration of the input dust, C_{in} (kg m^{-3}) [35, 36]:

$$\eta = \frac{C_{out}}{C_{in}} \tag{21}$$

A more widely used method is the Deutsch-Anderson model that is given by [37, 38]:

$$\eta = 1 - exp\left(-\frac{A}{Q}\omega_d\right) \tag{22}$$

Where A (m^2) is the surface area of the ESP, Q (m^{-3} s^{-1}) is the total gas flow rate, and ω_d (m s^{-1}) is the effective particle drift velocity that is calculated using the following relationship [39]:

$$\omega_d = \frac{q_p\left|\vec{E}\right|C_u}{3\pi d_p} \tag{23}$$

where q_p (C) is the particle charge, C_u is the Cunningham slip factor, μ is the viscosity of the flue gas (N S m^{-2}), and d_p (m) is the diameter of the particle.

3.6 Wire-plate ESP implementation on COMSOL multiphysics

One of the software packages that can be used for modeling the wire-plate ESP is COMSOL Multiphysics, a software package that employs FEA. This section aims to develop a comprehensive model that focuses on the electrostatic fields and particle behavior within the ESP without considering the airflow. For this reason, the model will utilize the electrostatics (es) module in COMSOL Multiphysics to simulate the electric field distribution between the wire and plate electrodes at different applied voltage levels. To successfully model a wire-plate ESP, the following steps will be employed: geometry creation, applying boundary conditions, calculating the threshold electric field, and performing mesh convergence test.

3.6.1 Geometry and material setup

The single-channel wire-plate ESP geometry, shown in **Figure 6**, comprises three *0.2 (mm)* wires spaced by $S = 40\ (mm)$. Additionally, there are two parallel plates with a width $W_p = 160\ (mm)$. The spacing is $d = 140\ (mm)$ between these plates. The chosen materials for this configuration are copper for both the discharge wires and collection plates, while the remaining volume is filled with air at standard atmospheric conditions.

3.6.2 Meshing and boundary conditions

The boundary conditions for the ESP model simulate the interaction between the electric field and charged particles, these are found in **Figure** 7 below. The specific conditions applied include:

- *Collecting plates*: A voltage potential of *0 V* was applied. Additionally, the charge density was set to $\rho_p = 0$.
- *Discharge wires*: The applied voltage of either $V_a = 20, 25, 30,$ *or* $35\ kV$ was used in this simulation. As the surface charge density is unknown, a threshold value for the electric field, E_t, is calculated, and an initial guess for the surface charge density ρ_0, is used. This value is adjusted iteratively using a parametric sweep until the electric field strength at each wire E_w that agrees with the threshold electric field, E_t, is achieved. The current value for ρ_0 is then used as ρ_w.
- *Inlet and outlet*: Zero diffusive flux condition, $\partial\rho/\partial n = 0$, was used to impose continuity.

The value of the threshold electric field can be calculated using Peek's law, with $\delta = 1$ (for 760 torr and *20°C*), and $r_w = 0.2$. Using these values, Eq. (6) yields:

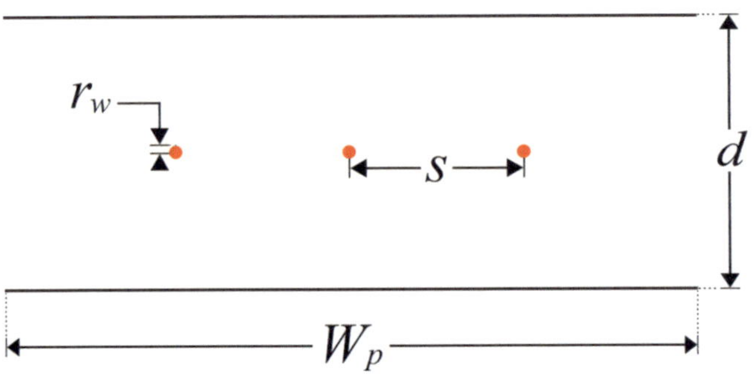

Figure 6.
Modeled geometry of the single-channel 3-wire ESP.

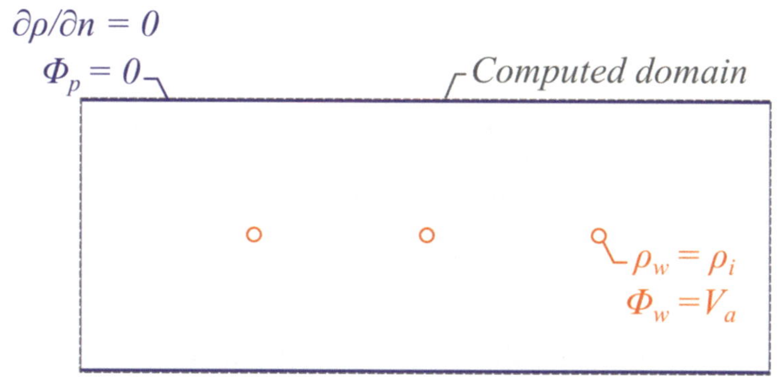

Figure 7.
Boundary conditions used in the simulation.

R_w (mm)	Parameter	Fine mesh	Finer mesh
0.2	$\rho_w\ (C\,m^{-2})$	1.44×10^{-4}	1.44×10^{-4}
	E_w (V m^{-1})	2.20×10^{6}	2.20×10^{6}
0.4	$\rho_w\ (C\,m^{-2})$	7.91×10^{-5}	8.15×10^{-5}
	E_w (V m^{-1})	2.20×10^{6}	2.20×10^{6}

Table 1.
Calculated values of ρ_w and the corresponding values of E_w.

$\overrightarrow{E_i} = 2.1959 \times 10^6\ (V\,m^{-1})$. Using the parametric sweep option in COMSOL Multiphysics, we get the following values for $\rho_w = 1.439 \times 10^{-4}\ (C\,m^{-2})$.

Initially, a predefined fine triangular mesh was employed. To establish the final element size, the surface charge density values for the wires, denoted as ρ_w, were calculated based on a specific threshold electric field, E_t. The current mesh size was retained when the values of ρ_w fell within a tolerance of $\pm 10^{-3}$ from the previous iteration. A stationary study was used to perform this simulation. The result of this procedure is shown in **Table 1** above.

3.6.3 Results

Corona discharge is a phenomenon characterized by the relationship between the current collected by conductive plates and the voltage applied to a wire, represented through an I–V (current-voltage) curve. This curve is essential for understanding the electrical behavior of corona discharge. The corona current can be computed by integrating the current density over the wire surface. At a given applied voltage value V_a, where ρ_w and E_w are the charge density and the electric field values at the wire, respectively, the current per unit length I is given by [40]:

$$I = \iint_S \vec{J} \cdot ds = 2\pi R_W \times \mu\, \rho_W E_W \tag{24}$$

The I–V curve, shown in **Figure 8**, has been validated with experimental results [40] of a wire-plate ESP.

Figure 9 illustrates the electrical potential distribution at various applied voltage levels. The potential distribution is measured along a vertical cutline extending from the center of the wire electrode to the collecting plate, visualizing the change of the electric potential in proximity to the electrodes. Notably, the electric potential reaches its maximum value at the centers of the wires, and decays to 0 when it reaches the collecting plates. As the applied voltage increases, the electric potential distribution becomes more pronounced, with higher voltages leading to a steeper gradient in electric potential. This steeper gradient can enhance the attraction of particles toward the collecting plate.

A similar procedure was employed in **Figure 10**, which illustrates the electric field strength at various discharge voltage levels. This gradient demonstrates how particles are influenced as they traverse the ESP. Specifically, increases in the applied voltage

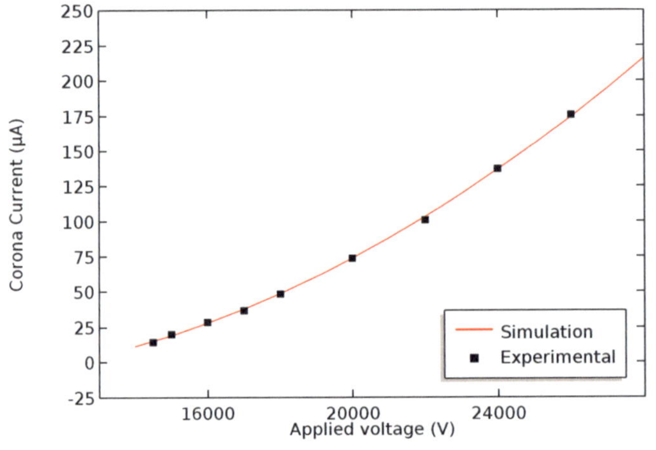

Figure 8.
Current-voltage characteristics at varying applied voltage levels.

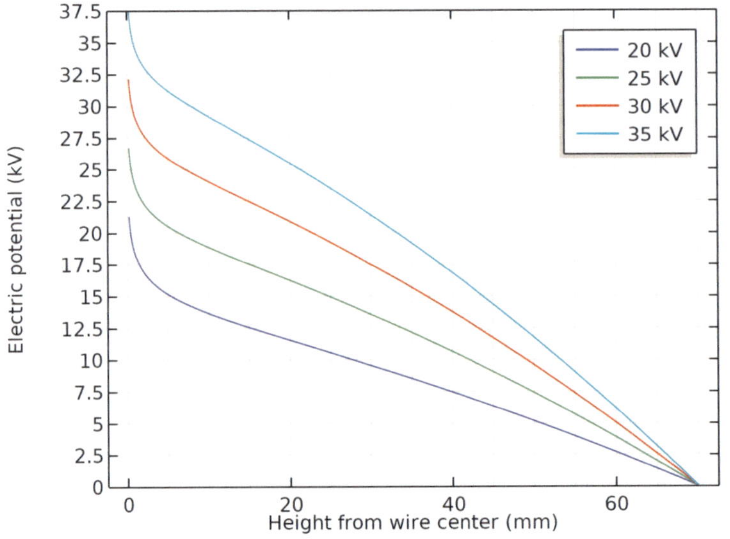

Figure 9.
Variation of electric potential distribution at different voltage levels.

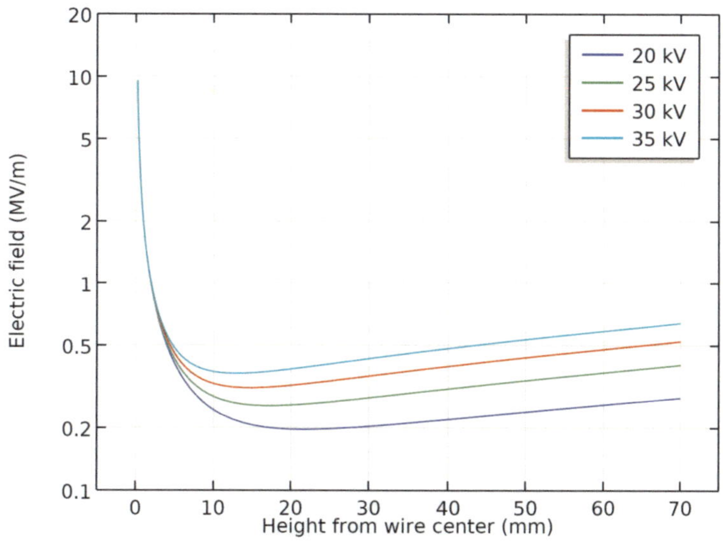

Figure 10.
Electric field strength at various discharge voltage levels.

correspond to a rise in the electric field strength, resulting in a more pronounced force exerted on the charged particles, thereby influencing their trajectory and subsequent collection on the plates. It is noteworthy that there is a sudden decay in electric field strength after the electrode, which can be attributed to the rapid dispersion of electric field lines as they propagate away from the concentrated area near the discharge electrode.

The distribution of ion charge density, illustrated in **Figure 11**, demonstrates how charges accumulate and are distributed in response to the applied electric field. As

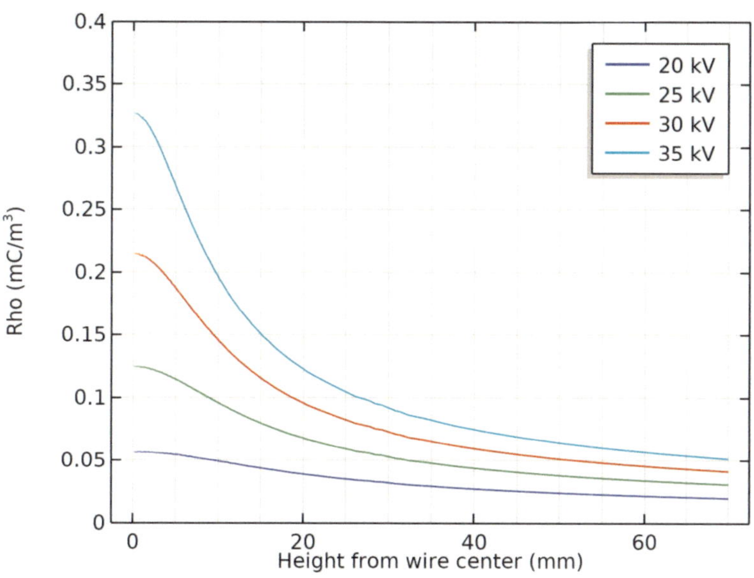

Figure 11.
Ion charge density distribution at various discharge voltage levels.

voltage increases, ion charge density rises, leading to greater charged particle accumulation. The increased charge density facilitates a more effective particle migration, thereby improving the overall efficiency of the ESP. The ion charge density is relatively uniform at lower voltage levels, as the voltage increases, however, the charge density becomes more concentrated near the discharge electrode, reflecting the enhanced ionization and particle charging. There is a sudden decline in ion charge density immediately after the discharge wire as particles approaching the grounded plate experience reduced electric field strength, diminishing charge retention.

4. Industrial considerations of ESPs

ESPs play a crucial role in controlling particulate emissions from industrial processes. The efficiency of an ESP in collecting particulate matter is influenced by several operational and design parameters. This section discusses these parameters affecting precipitation efficiency, including inlet flow velocity, power consumption, specific collection area, plate-to-plate distance, corona power ratio, and the aspect ratio of the ESP.

Understanding these parameters is essential for optimizing the performance of ESPs. By exploring these factors, we aim to provide a comprehensive overview of how they interact and contribute to the overall effectiveness of ESPs in an industrial context.

4.1 Power consumption

The relationship between input power and precipitation efficiency is complex; greater input power generally leads to higher efficiency up to a specific threshold. Beyond this limit, increasing power does not yield significant improvements in collection efficiency, indicating the presence of diminishing returns [10]. This phenomenon underscores the importance of optimizing power consumption to achieve effective particulate removal without incurring unnecessary operational costs.

The corona power ratio, denoted as P_c, and expressed in units of [W·s/m^3], is a commonly used metric for assessing the power consumption of ESPs in real-world scenarios from an economic point of view [12, 41]. It is defined as the ratio of the total power consumed by the ESP, P, to the gas flow rate, Q, through the system [42].

$$P_c = {}^{P}\!/_{Q} \tag{25}$$

The total power consumed, P, is calculated as the product of the current, I, and the voltage, V, applied to the ESP:

$$P = I \cdot V \tag{26}$$

4.2 Inlet flow velocity

Variations in inlet flow velocity can have a profound impact on particle charging times and, consequently, the overall precipitation efficiency of the ESP. This is because higher velocities reduce the residence time of the particles within the electric field, limiting the time available for effective charging [38, 43]. Conversely, lower inlet velocities generally yield higher precipitation efficiencies by allowing for longer particle charging times [44].

In addition, this effect is further compounded by the localized high-velocity zones that form within the ESP, as these zones lead to reduced particle residence times and thus decreased charging efficiency in those regions, ultimately compromising the overall performance of the ESP [45]. To maintain consistent inlet velocities, proper design considerations such as adjusting the profile, and the cross-sectional area of the inlet duct can be implemented to ensure uniform velocity distribution at the ESP inlet [44].

4.3 Specific collection area

The specific collection area (SCA) of an ESP measures the total surface area of collecting electrodes over the volumetric flow of the flue gas [46]. The SCA directly influences the collection efficiency of the ESP [47], as it determines the available surface area for particulate capture. A larger SCA typically correlates with improved efficiency in removing particles from the gas stream, as it provides more area for the charged particles to adhere to the collecting surfaces. However, it is essential to balance the SCA with cost considerations, as larger collection areas can lead to increased capital and operational expenses. This is determined through the sizing procedure [13], which determines the total area of collection surfaces for the specified performance.

4.4 Plate-to-plate distance

Plate-to-plate spacing is an important parameter for evaluating the performance of an ESP [35, 36]. The collection efficiency is generally higher when the spacing between subsequent collection plates is narrower for particles in the micrometer range, assuming a constant current density between the collecting electrodes [19, 36, 42, 48]. Conversely, wider spacing can lead to higher power consumption due to the reduced electric field strength, which requires higher voltages to maintain the necessary field strength for the same collection efficiency [48]. The increased power consumption can offset the potential benefits of wider spacing, such as reduced installation costs. In contrast, in nanoparticles, where the charging is caused by diffusion, greater spacing promotes higher collection efficiency [49].

4.5 Discharge electrodes radii and spacing

The increase in the radius of the discharge electrode leads to improved collection efficiency provided that the average current density at the collection plates remains constant in wire-plate ESPs [12]. This relationship is reversed when it is evaluated at a constant electric field [42]. This phenomenon occurs due to an increase in the onset corona voltage and a reduction in the space charge effect [36].

Additionally, it is important to consider that smaller discharge wire radii facilitate an earlier onset of corona discharge [50], which is particularly advantageous in applications requiring low voltage levels. The optimal wire-to-wire spacing in terms of precipitation efficiency depends on the specific operating conditions. While a definitive relationship between spacing and performance has yet to be established, several models attempted to address this [51, 52].

5. Conclusion

Electrostatic precipitators remain a crucial technology for controlling particulate emissions in various industrial settings. Their ability to efficiently capture a wide range of particle sizes makes them necessary in meeting increasingly stringent environmental regulations.

In terms of design, wire-plate, transverse plate, and wire-cylinder configurations each offer unique advantages, tailored toward specific needs. Similarly, single-stage and multi-stage ESPs present trade-offs in terms of simplicity, collection efficiency, and resilience to variable operating conditions.

Understanding the fundamental principles of ESP operation, from corona discharge to particle charging and migration, is essential for optimizing their performance. Computational modeling approaches, such as those employing FEA, provide valuable insights into the complex interactions between electric fields, airflow patterns, and particle dynamics within ESPs. These models enable the exploration of design variations.

In the industrial context, several key parameters must be considered to maximize the efficiency and cost-effectiveness of ESPs. Power consumption, inlet flow velocity, specific collection area, plate-to-plate distance, and aspect ratio all play crucial roles in determining the overall performance of the system. By striking a balance between these factors, engineers can design ESPs that deliver high collection efficiencies while minimizing operational costs.

As environmental concerns continue to drive the need for effective particulate matter control, the role of ESPs will only grow in importance.

Author details

Mohamed Badran* and Abdallah Alkhaiyat
Department of Mechanical Engineering, The American University in Cairo, New Cairo, Egypt

*Address all correspondence to: mobadran@aucegypt.edu

References

[1] Zelikoff JT, Chen LC, Cohen MD, Fang K, Gordon T, Li Y, et al. Effects of inhaled ambient particulate matter on pulmonary antimicrobial immune defense. Inhalation Toxicology. 2003; **15**(2):131-150

[2] Mizuno A. Electrostatic precipitation. IEEE Transactions on Dielectrics and Electrical Insulation. 2000;**7**(5):615-624

[3] Pope CA. Review: Epidemiological basis for particulate air pollution health standards. Aerosol Science and Technology. 2000;**32**(1):4-14

[4] Molchanov O, Krpec K, Horák J. Electrostatic precipitation as a method to control the emissions of particulate matter from small-scale combustion units. Journal of Cleaner Production. 2020;**246**:119022

[5] Guo BY, Yang D, Su YB, Yu AB. Process modelling of low temperature electrostatic precipitators. Powder Technology. 2017;**314**:567-576

[6] Jena MC, Mishra SK, Moharana HS. Effect of particle size on collection efficiency of ESP and RABH: A case study. Aerosol Science and Engineering. 2019;**3**(3):75-87

[7] Miller BG. Emissions control strategies for power plants. In: Coal Energy Systems [Internet]. Elsevier; 2005. pp. 283-392. [cited 2024 Oct 31]. Available from: https://linkinghub.elsevier.com/retrieve/pii/B9780124974517500061

[8] Sicard P, Agathokleous E, De Marco A, Paoletti E, Calatayud V. Urban population exposure to air pollution in Europe over the last decades. Environmental Sciences Europe. 2021; **33**(1):28

[9] Amann M, Kiesewetter G, Schöpp W, Klimont Z, Winiwarter W, Cofala J, et al. Reducing global air pollution: The scope for further policy interventions. Philosophical Transactions of the Royal Society A. 2020;**378**(2183):20190331

[10] Badran M, Mansour A. Electrostatic Precipitators: Impact of Design Parameters on Collection Efficiency [Internet]. Eliva Press; 2023. 65 p. Available from: https://www.elivapress.com/en/book/book-6077765264/

[11] White HJ. Fifty years of electrostatic precipitation. Journal of the Air Pollution Control Association. 1957;**7**(3):166-177

[12] Badran M, Mansour AM. Evaluating performance indices of electrostatic precipitators. Energies. 2022;**15**(18):6647

[13] Turner JH, Lawless PA, Yamamoto T, Coy DW, Greiner GP, McKenna JD, et al. Sizing and costing of electrostatic precipitators: Part I. Sizing considerations. JAPCA. 1988;**38**(4): 458-471

[14] Yuen AWL. Collector Current Density and Dust Collection in Wire-Plate Electrostatic Precipitators [Internet]. Sydney: UNSW; 2006. [cited 2024 Nov 6]. Available from: http://hdl.handle.net/1959.4/28274

[15] Yi CW, Yin T, Huang HQ, Du YS, Yi RJ. Study on new laboratory scale transverse plate electrostatic precipitator collecting fly ash. AMM. 2013;**433-435**: 2260-2264

[16] Mermigkas A. Micro-Electrostatic Precipitation for Air Treatment [Internet]. University of Strathclyde; 2016. [cited 2024 Nov 6]. Available

from: https://stax.strath.ac.uk/concern/theses/3f4625492

[17] Cao R, Tan H, Xiong Y, Mikulčić H, Vujanović M, Wang X, et al. Improving the removal of particles and trace elements from coal-fired power plants by combining a wet phase transition agglomerator with wet electrostatic precipitator. Journal of Cleaner Production. 2017;**161**:1459-1465

[18] Zhu Y, Chen C, Shi J, Shangguan W. Enhancement of air purification by unique W-plate structure in two-stage electrostatic precipitator: A novel design for efficient capture of fine particles. Advanced Powder Technology. 2020; **31**(4):1643-1658

[19] Gao M, Zhu Y, Yao X, Shi J, Shangguan W. Dust removal performance of two-stage electrostatic precipitators and its influencing factors. Powder Technology. 2019;**348**:13-23

[20] Parker KR. Applied Electrostatic Precipitation. London: Blackie; 1997

[21] Farnoosh N, Adamiak K, Castle GSP. 3-D numerical analysis of EHD turbulent flow and mono-disperse charged particle transport and collection in a wire-plate ESP. Journal of Electrostatics. 2010;**68**(6):513-522

[22] Dastoori K, Makin B, Kolhe M, Des-Roseaux M, Conneely M. CFD modelling of flue gas particulates in a biomass fired stove with electrostatic precipitation. Journal of Electrostatics. 2013;**71**(3): 351-356

[23] Yamamoto T, Velkoff HR. Electrohydrodynamics in an electrostatic precipitator. Journal of Fluid Mechanics. 1981;**108**:1-18

[24] Cristina S, Dinelli G, Feliziani M. Numerical computation of corona space charge and V-I characteristic in dc electrostatic precipitators. IEEE Transactions on Industry Applications. 1991;**27**:7

[25] Peek F. Dielectric Phenomena in High-Voltage Engineering. New York: McGrawHill; 1929

[26] Park JW, Kim C, Park J, Hwang J. Computational fluid dynamic modelling of particle charging and collection in a wire-to-plate type single-stage electrostatic precipitator. Aerosol and Air Quality Research. 2018;**18**(3):590-601

[27] Gerhart PM, Gerhart AL, Hochstein JI. Fundamentals of Fluid Mechanics. 9th ed. Hoboken: Wiley; 2020

[28] Skodras G, Kaldis SP, Sofialidis D, Faltsi O, Grammelis P, Sakellaropoulos GP. Particulate removal via electrostatic precipitators—CFD simulation. Fuel Processing Technology. 2006;**87**(7):623-631

[29] Dastoori K, Kolhe M, Mallard C, Makin B. Electrostatic precipitation in a small scale wood combustion furnace. Journal of Electrostatics. 2011;**69**(5): 466-472

[30] Parry M, Domina S. Field Theory for Engineering. D. New York: Van Nostrand Co.; 1961

[31] Graebel WP. Engineering Fluid Mechanics. New York: Taylor & Francis; 2001. 676 p

[32] Nair SA, Yan K, Pemen AJM, Winands GJJ, Van Gompel FM, Van Leuken HEM, et al. A high-temperature pulsed corona plasma system for fuel gas cleaning. Journal of Electrostatics. 2004; **61**(2):117-127

[33] Coulson JM, Richardson JF. Particle Technology and Separation Processes.

Woburn: Butterworth Heinemann; 2002. (Chemical Engineering; vol. 2)

[34] Jaworek A, Krupa A, Czech T. Modern electrostatic devices and methods for exhaust gas cleaning: A brief review. Journal of Electrostatics. 2007; **65**(3):133-155

[35] Navarrete B, Cañadas L, Cortés V, Salvador L, Galindo J. Influence of plate spacing and ash resistivity on the efficiency of electrostatic precipitators. Journal of Electrostatics. 1997;**39**(1): 65-81

[36] Abdel-Sattar S. Influence of geometrical parameters upon electrostatic precipitator efficiency. COMPEL. 1991;**10**(1):27-43

[37] Gouri R, Zouzou N, Tilmatine A, Moreau E, Dascalescu L. Collection efficiency of submicrometre particles using single and double DBD in a wire-to-square tube ESP. Journal of Physics D: Applied Physics. 2011;**44**(49):495201

[38] Chang Y, Jia P, Shi L, Xiang X. Corona discharging and particle collection of bipolar transverse plate ESP. Journal of Electrostatics. 2018;**96**: 104-110

[39] Neimarlija N, Demirdžić I, Muzaferija S. Numerical method for calculation of two-phase electrohydrodynamic flows in electrostatic precipitators. Numerical Heat Transfer, Part A: Applications. 2011;**59**(5):321-348

[40] Kasdi A. Computation and measurement of corona current density and V–I characteristics in wires-to-plates electrostatic precipitator. Journal of Electrostatics. 2016;**81**:1-8

[41] Ning Z, Podlinski J, Shen X, Li S, Wang S, Han P, et al. Electrode geometry optimization in wire-plate electrostatic precipitator and its impact on collection efficiency. Journal of Electrostatics. 2016;**80**:76-84

[42] Chang CL, Bai H. Effects of some geometric parameters on the electrostatic precipitator efficiency at different operation indexes. Aerosol Science and Technology. 2000;**33**(3): 228-238

[43] Xiang X, Chang Y, Nie Y. Investigation of the performance of bipolar transverse plate ESP in the sintering flue control. Journal of Electrostatics. 2015;**76**:18-23

[44] Haque SME, Rasul MG, Khan MMK, Deev AV, Subaschandar N. Influence of the inlet velocity profiles on the prediction of velocity distribution inside an electrostatic precipitator. Experimental Thermal and Fluid Science. 2009;**33**(2):322-328

[45] de Oliveira AE, Guerra VG. Influence of particle concentration and residence time on the efficiency of nanoparticulate collection by electrostatic precipitation. Journal of Electrostatics. 2018;**96**:1-9

[46] Reynolds JP, Theodore L, Marino J. Calculating collection efficiencies for electrostatic precipitators. Journal of the Air Pollution Control Association. 1975; **25**(6):610-616

[47] Li S, Zhang S, Pan W, Long Z, Yu T. Experimental and theoretical study of the collection efficiency of the two-stage electrostatic precipitator. Powder Technology. 2019;**356**:1-10

[48] Mansour A, Badran M. Modelling and simulating the effect of the different geometric parameters on voltage-current characteristics for wire-plate electrostatic precipitator with different

collector configurations. In: 2020 21st International Conference on Thermal, Mechanical and Multi-Physics Simulation and Experiments in Microelectronics and Microsystems (EuroSimE) [Internet]. Cracow, Poland: IEEE; 2020. pp. 1-8. [cited 2020 Aug 21]. Available from: https://ieeexplore.ieee.org/document/9152744/

[49] De Aquino LF, Guerra VG. Influence of wire spacing and plate spacing on electrostatic precipitation of nanoparticles: An approach involving electrostatic shielding and diffusion charging. Particuology. 2023;**80**:127-139

[50] Conesa AJ, Sánchez M, León M, Cabrera Á. Some geometrical and electrical aspects on the wire-to-cylinder corona discharge. Journal of Electrostatics. 2019;**100**:103355

[51] Zhibin Z, Guoquan Z. New model of electrostatic precipitation efficiency accounting for turbulent mixing. Journal of Aerosol Science. 1992;**23**(2):115-121

[52] Kim SH, Lee KW. Experimental study of electrostatic precipitator performance and comparison with existing theoretical prediction models. Journal of Electrostatics. 1999;**48**(1):3-25